ENERGY EMPIRES IN COLLISION

ENERGY-EMPIRES IN COLLISION

The Green versus Black Struggle for our Energy Future

by

David Howell

G

GILGAMESH
PUBLISHING

By the same author:

Freedom and Capital: Prospects for the Property-owning Democracy (Blackwell, 1981).

Blind Victory (Hamish Hamilton, 1986).

The Edge of Now: New Questions for Democracy in a Precarious World (Macmillan, 2001).

Out of the Energy Labyrinth: Uniting Energy and the Environment to Avert Catastrophe co-authored with Carole Nakhle (I. B. Tauris, 2007).

Old Links and New Ties: Power and Persuasion in an Age of Networks (I. B. Tauris, 2013).

Energy Empires in Collision

Published by Gilgamesh Publishing in 2017
Email: info@gilgamesh-publishing.co.uk
www.gilgamesh-publishing.co.uk

ISBN 978-1-908531-773

CIP Data: A catalogue for this book is available from the British Library

Dedication

To my wife, Davina

I have made a gift to mine enemies. I have written a book.

Anon.

To achieve anything one has to rise above the noise of politics.

HRH Prince Hassan bin Talal of Jordan

CONTENTS

PREFACE 13

PROLOGUE 21

Chapter One

ENERGY EMPIRES IN COLLISION 27

The divided world energy scene – the entrenched empire of oil, gas, and coal versus the 'irresistible' rise of green power – the judgement of Paris – prospects for global energy transformation – oil, gas, and coal abundance – can fossil fuels be phased out? – growing hostility between the rivals but some common ground – neither side can win – how cheap oil disrupts – how renewable energy costs rise – the public reaction – misleading forecasts – forces at work beyond any government's control – Lord Stern asks, 'Why are we waiting?' – Lord Stern's question answered.

Chapter Two

THE CRADLE ROCKED 54

The Middle Eastern oil kingdoms in crisis – the triple threat to oil and gas riches – encroaching regional violence and instability – swelling non-OPEC oil and gas production – opinion and governments turn against fossil fuels – the old assumptions of Middle Eastern energy dominance fade – the post-2008 shocks – oil and gas demand weakens – the impact of Asian growth and China – has OPEC lost control? – the clever Emiratis who ride both horses.

Chapter Three

FRESH UPHEAVALS 69

The power shift to Asia – main energy consumption growth outside the West – climate concerns mount – governments and popular opinion against oil – green subsidies and hydrocarbon penalties – new uncertainties for investors – the 'third earthquake' of hydraulic fracturing and the shale revolution.

Chapter Four

LOW-CARBON PARADISE POSTPONED? 76

The trilemma failures: 1. Global carbon emissions set to continue rising – how gas cuts carbon – American success but world failure – targets certain to be missed – the dangers of dogma – nuclear power hopes dashed with German opt-out – coal prevails in Asia but also in Europe – the limitations of carbon capture and storage – Europe's 'example' strategy dead.

Chapter Five

GOODBYE CHEAP ENERGY 102

The trilemma failures: 2. Europe goes for expensive energy – the true costs assessed – the backlash of public protest – no mercy for the poor – the billions who lack cheap power – an odd political stance on the Left – the real low-carbon path.

Chapter Six

FLICKERING LIGHTS 109

The trilemma failures: 3. Unreliable energy – self-sufficiency not always safest – green ambitions weaken network security – the struggle to persuade new investment in new electricity-generating capacity – the expense of carbon capture and storage – other technologies might help.

Chapter Seven

EUROPEAN POLICY: ENERGY DISUNION AND THE UPSIDE-DOWN STRATEGY 116

How EU policy blunders have undermined both energy and climate strategies – Europe miscalculates – all fossil fuels initially opposed – the Russian wake-up call – the Fukushima impact – other avenues blocked – no real European energy market – ambivalence between security and decarbonization.

Chapter Eight

RUSSIAN DOLLS, GERMAN MUDDLES, POLISH DISMAY 130

What does Russia really want? – Central Europe's Russian gas dependence – the Crimea-Ukraine 'shock' – other European gas sources considered – Russia looks east.

The German energy policy muddle – Energiewende and the massive expansion of the dirtiest kinds of coal-burning – damage to German industry.

Polish dependence on coal – shale hopes dashed – Polish clean coal ambitions – hostility to EU renewables plans – an example of EU rigidity.

Chapter Nine
THE COALITION LEGACY: A VERY BRITISH FIASCO 145

Frailty of the UK energy system – the total trilemma policy failure – the blame game – a case for the defence – National Grid and keeping the lights on – coal dismissed but coal persists – picking the renewables winners – the capacity payments muddle – nuclear renaissance and the Hinkley C problem – a change of direction begins – the constraints of EU policy.

Chapter Ten
ESCAPE FROM THE LABYRINTH 180

The continuing contradictions and stupendous errors of EU energy policy – the possible escape routes – Caspian pipeline possibilities – weak shale prospects – North Sea decline – a policy that might work – switching resources to green energy research and technology – creating a pipeline grid and energy market – Russian gas to Europe will continue – how Britain and Germany should be making a different contribution to meeting climate challenges – prospects for much cheaper green power.

Chapter Eleven

THE PRICE 207

How the crude oil price decides everything – the predicted upward spiral in price fails to appear – oil price volatility throughout history – price forecasts usually wrong – price collapse and the Middle Eastern producers – the link with the jihadis – world oil supply abundance – OPEC and the lack of Saudi policy choice – false hopes for non-OPEC 'responsibility' – estimated break-even points for oil-producing countries – the geopolitical consequences of weak oil prices – are lower oil prices better for both producers and consumers?

Chapter Twelve

BUT WHEN WILL GREEN MEAN CHEAPER? 235

The price prospects for renewables – how much dare governments charge consumers? – when can renewable energy sources become commercial and competitive? –the encouraging American experience – research and the Apollo Programme.

Chapter Thirteen
BITING THE ATOM: TO BUILD OR NOT TO BUILD 245
Nuclear power stations: the endless loop of doubt and decision – the lessons of Fukushima – facts about worldwide nuclear power station construction – is there a renaissance? – Germany and Japan go opposite ways – Hinkley, Hualong, and the Chinese nuclear entrée into Europe – is Hinkley C the last of the line? – prospects for small modular reactors – a race against time.

Chapter Fourteen
THE CHINA SYNDROME 264
Another Chinese visit – all energy roads lead to China – China, the dominant energy 'player'- China, the biggest coal producer and coal importer – Chinese contradictions – Chinese industrialization path different from the West's – China's worldwide reach in Africa, in Asia, in Europe – emphasis of One Belt, One Road on oil and gas development – green target rhetoric versus hydrocarbon plans and realities.

Chapter Fifteen
THE TOMORROW WE DESERVE 284
The pathway to a better energy future – the pathway blocked – the recognition of error, a first step –

zealotry, the enemy of progress – technology, the real driver of change – harnessing green capitalism – the unavoidable world expansion of fossil fuels – the absolute imperative of cost-effective carbon capture – cooperation between green and hydrocarbons worlds essential – turning energy abundance into cheap, reliable, low-carbon energy – the next ten years – the unending energy revolution – will it ever be dull?

******

PREFACE

The first (hardback) edition of this book contained a health warning. The warning was that it was an unfinished work, and indeed could never be finished. This was due to the headlong pace of technological change in the energy world which was bound to overtake, and often invalidate, all written analyses of energy and energy-related issues long before they could be printed and published.

So it turned out with *Empires in Collision*. Published in March 2016, it could just about catch up to the point where the oil price was clearly set to stay low for a long while to come, where the full cost horrors of the UK's giant Hinkley Point nuclear reactor were becoming starkly apparent and where the EU's mishandling of its energy and climate policies was becoming more evident by the day.

What it did not foresee – and nor did most people - was that, in the months immediately following publication, the UK would vote to leave the European Union for good and that Donald Trump would succeed in his bid for the White House. Both these events have had a Richter-scale impact on world energy, creating the demand for updated comment and assessment.

In London, the Prime Minister and the Chancellor were swept away, and with them the Whitehall Department of Energy and Climate Change – which many regarded as having been far more concerned with climate change objectives than with safeguarding energy supplies. In Washington, the President-elect, Donald Trump, chose as new head of the Environmental Protection Agency, Scott Pruitt, a long-standing and powerful critic of the climate change case and of measures to combat climate change.

He went on to nominate Rex Tillerson, the ex-chief of Exxon and a champion of the oil and gas worlds, as Secretary of State and America's face to the world.

As the Trump administration gets underway these appointments and numerous Trump statements – to the effect that the Paris Climate Treaty of November 2015 should be scrapped – can leave no-one in any doubt that the 'collision' between fossil fuel interests and the demands of low carbon campaigners – the black versus green struggle at the centre of this book – has been vastly intensified.

Whether that means a contraction of all lower carbon policies or just smarter (i.e. more effective and less costly) climate policies remains to be seen. Bjørn Lomborg argues with some force that the overall invigoration of all energy sources and power production which the President-elect has promised could well lead not to abandonment of climate concerns but to a more effective range of policies, costing less per tonne of carbon reduced and ultimately less burdensome on consumers and less threatening to reliable power supplies - a better-than-Paris performance.

This would be much in line with the contention of *Empires in Collision* – that instead of war between the oil industry and the green cause, a unity of purpose on common ground gives by far the best hope of achieving all the goals: cheap power, and reliable power supplies as well as a lower carbon world.

As for the Brexit decision, some argued at first that neither UK energy and climate policies, nor those throughout the rest of Continental Europe, would be much affected. But here we come up against the second iron law of the digital age, namely that the more we become connected, instantly and continuously, the more vulnerable to the actions, events and policies of everybody else we become. This now applies generally in all our lives and in all international affairs, but in the energy field especially.

Because of mismanagement, the UK is going to be relying over the next decade for up to a quarter of its daily power supplies on interconnectors with European neighbours. Not all of them will be in the EU – Norway and Iceland will be big suppliers through interconnector cables and systems yet to be constructed . But so will Denmark, France (as it already is), Belgium and the Republic of Ireland.

Managing daily, indeed hourly, connections with these neighbours will require close and very good cooperation, come what may in any negotiations. There will also be a need to re-write the Euratom Treaty which governs the regulation of Europe's nuclear plants and the safe handling of uranium supplies for enrichment.

Meanwhile, the whole world energy scene has continued to demonstrate its own versions of Moore's law (Gordon Moore's prediction that that the power of microchips would double roughly every two years), along with dozens of other powerful innovations and new trends. Quite aside from what may be triggered by the political upheavals, the energy cosmos continues to undergo deep changes, sending ripples of change through our societies and daily lives.

Thomas Friedman, in his latest book (*Thankyou for Being Late*, Allen Lane 2016) emphasizes the point that change is not coming at a sedate linear pace. On the contrary, it is accelerating – a hurtling, expanding stream of new technologies, moving faster than people can adjust and faster than regulators or law-makers can keep pace with.

Who would have thought as little as five years ago that cargoes of American shale gas would have begun to arrive in Europe? Or that OPEC would be floundering and probably sinking in a sea of surplus oil inventories? Thanks to new technologies, fracking methods have become far more efficient, and the US shale sector consequently is much more resilient in the face of low oil and gas prices. Moreover, the Texans have found a new oil field of Saudi proportions, set to meet US demand for years to come. Turkey has made it up with Russia, allowing Russian plans to pipe gas through Turkish grids to Southern Europe.

New, cheaper and cleverer ideas have begun to circulate for capturing carbon in fossil fuel burning. Battery storage - the Holy Grail for intermittent wind and solar power – has made further leaps, as has battery technology in the automotive industry. Tesla Motors has come to market with its revolutionary electric cars and has set up a smart UK showroom on the Hogarth Roundabout at Chiswick to dazzle British customers – or at least those with deep pockets.

On another big low carbon front, the final go-ahead decision on EDF's Hinkley Point nuclear plant has been taken, after much wobbling, alongside an acceleration of interest and effort in the field of small modular reactors. All across the renewable energy field, costs are falling month by month – solar energy in particular, but also wind power, tidal power, and stored power, all driven by ceaseless ingenuity and innovation.

At the demand end of the energy chain, a cascade of new methods and devices for increasing energy consumption efficiency has continued, many reaching commercial levels and becoming fully available to the consumer. And the peculiarly British muddle over energy supply policy has continued as before, with efforts to meet peak winter demand resulting in wafer thin electricity supply safety margins and in millions of pounds being paid to coal burning generating stations, despite plans to shut them by 2026. Diesel operators have also been called in – all at soaring cost to consumers – a true Alice-in-Wonderful example of upside-down thinking, equalled only in Germany where vast subsidy expenditure in line with the ambitious Energiewende programme has resulted in more dirty coal being burnt, not less, and in higher, not lower CO2 emissions.

Yet amidst all the confusion and argument, the message of *Empires in Collision* during this period of continuing hectic change and political upheaval remains essentially the same. If the great energy transition is to succeed, if carbon emission levels are to be anywhere near curbed, if billions of people in developing economies are to have the chance of cheap power, if in the more developed world, reliable power supplies are to be secured, the key will not be the war between the green campaigners and the world's energy industries, but clever and highly constructive cooperation. It is this that will matter above all else.

April 2017

PROLOGUE

Energy Entry into the Department at the Centre of Everything

May 7th 1979. The wind moans through the ill-fitting windows.

I am being shown into a large hospital-white room with some long, squashy sofas and chairs that look as though they have seen better times, a repro mahogany table and about twelve chairs and an enormous desk at the end. We are in the palatial but shabby offices along Millbank once occupied by the old ICI, now a major government department.

The Permanent Secretary is politely smiling, waiting to receive me. I am taking over the department at the centre of everything. This is the department of oil shocks, the Shah having just fallen, the department of militant miners and Arthur Scargill both itching to have a go at the new Tory Government, the department of colossal investment programmes in mammoth, nationalized industries, the department of booming North Sea oil and North Sea licensing of blocks to be explored, with a state oil company owning and trading one of the largest volumes of oil on the planet, the department that has to keep alongside rising OPEC power, whose Sheikh Yamani and Sheikh Khalifa Al-Sabah would soon be pressing me over sandwiches at the Belfry off Belgrave Square (now Mossimans) to make Britain join OPEC – a fruitless quest.

It is the department of nuclear energy, of the vast British Gas empire, under its formidable boss, Sir Denis Rooke, of the Central Electricity Generating Board and all its area electricity companies, and of the National Coal Board, the department of relations with all the international oil companies. It is the department of the Gulf States and of Norway, the department which owns 51% of BP, the department of global energy turmoil, the department of soaring oil and gas prices and threatened oil shortages, rocking the world's economies, and of crisis meetings of the International Energy Agency in Paris. ('You will be chairing that next, Minister', so I'm told.)

It is the department promoting the growth of one of the world's biggest oil and gas industry supply chains, locked in quarrels about subsidies with Brussels, talking daily on the telephone with Jim Schlesinger (the US Energy Secretary) and other energy ministers round the globe, all struggling with leaping oil prices and OPEC threats of cut-offs. It is the department trying to encourage energy conservation, unfamiliar then, but now the biggest resource of all, with the energetic help of Bernard Ingham looking after that division (before he goes to off to greater things at No. 10). It is the department under daily bombardment from the Treasury to sell this, cut that, squeeze out more revenues from the other, the goose to be plucked. It is the department to which the new Prime Minister, Margaret Thatcher, takes an instant dislike, demanding that the Permanent Secretary, Sir Jack Rampton, whom I find very pleasant and helpful, be removed.

It is, in short, a department of Soviet proportions supposedly presiding over a huge socialized sector, about a quarter of British industry, an employer of millions of people, a consumer of billions of pounds and generator of billions more – a world which is all coming to the end of its time. We are poised right on the pivotal moment of the twentieth century as state mega-ownership and centralization finally choke themselves to death and the era of decentralization, denationalization, privatization, and the rising market state is about to begin.

Finally, it is the department which has to be explained almost daily to an angry and bewildered Parliament, with its Secretary of State described by Hugh Fraser MP as ruling the biggest ministerial empire of all time. Beyond Parliament, there are armies of even angrier motorists, indignant companies, and furious consumers. At the end of the first week of Thatcher's Government, the pump price went up by a then deeply painful 8p a gallon. 'Not a very good start', said Mr. Lloyd at the garage that weekend as he filled up the car at my in-laws' Welsh home – 'not a very good start at all'. There will be much more to come.

Just now in May 1979, this is also a demoralized department much as Tony Benn left it a few days before. I have inherited a superb chief private secretary direct from him who of course is suitably tight-lipped about what had gone on but discreetly gives me a flavour of the joyful madness in the last days of the previous regime. It seems that by the end of the Labour Government, Tony Benn was pursuing a sort of UDI policy within Whitehall. He simply refused to attend cabinet meetings and carried on independently with his sprawling empire. In the private office desk draw is a diamond-studded Rolex watch, a gift from a Saudi minister. By special dispensation, it has been allowed to be kept in the Department instead of being handed in to the government store, as all such gifts are supposed to be. This is so that the minister could be seen politely wearing it when the next Saudi prince or minister steps through the door.

Nothing like this immense departmental empire, with the fate of the whole government and economy on its plate, and almost with its own foreign policy, would ever exist again. It was of course unmanageable, uncontrollable, impossible, and fascinating. Decades ahead, entirely new energy and energy-connected issues would unfold. Issues like climate change and carbon dioxide emissions, never mentioned then, would come demandingly to the fore. Powerful, new green empires would rise to challenge the long-standing global domination of the hydrocarbons – coal, oil, and gas. But all that lay far ahead.

Then, as today, there was an abundance of energy but also an abundance of energy problems and a mountain of risks. That was, and remains, the paradox of energy. While the empires struggle, there is ample supply of energy for all and every one of us.

But in the meantime, 'Welcome to the Department of Energy, Secretary of State, you will need this. It is the key to your private toilet down the corridor.'

******

CHAPTER ONE

Energy Empires in Collision

The divided world energy scene – the entrenched empire of oil, gas, and coal versus the 'irresistible' rise of green power – the judgement of Paris – prospects for global energy transformation – oil, gas, and coal abundance – can fossil fuels be phased out? – growing hostility between the rivals but some common ground – neither side can win – how cheap oil disrupts – how renewable energy costs rise – the public reaction – misleading forecasts – forces at work beyond any government's control – Lord Stern asks, 'Why are we waiting?' – Lord Stern's question answered.

Today we live in two energy worlds, telling us two stories. One is the world of big oil and gas. The other is the world of green and low-carbon hopes and ambitions.

One of them is the long reigning and colossal empire of fossil fuels – oil, gas, and 'king' coal – the massive and intricate hydrocarbon order that has driven industry and ruled the earth for well over a century and whose backers are even now investing trillions more in projects confidently planned to last another half-century at least. It is the universe of giant oil and gas corporations, state-owned and private, of bottomless Middle Eastern and Emirati wealth, of Texan billionaires and Russian oligarchs, of booming gas and oil fracking, of thousands of miles of pipelines, of sprawling refinery complexes, of gas- and coal-burning power stations, of global coal mining and trade expanding on an unprecedented scale, of epic struggles for access to Arctic resources, of giant deep-sea oil platforms, of the world's biggest ships, of petrochemicals, and, of course, of the universal petrol or diesel pump, gas stoves and canisters everywhere, of lit-up cities and homes, of thousands of oil-based products found in the wheels of every industry and most machines. It is an all-pervasive world not only of plenty but also one which some believe to be leading us into an age of a persisting oil and gas glut.

The other newer empire, rising to challenge all fossil fuels, is the alternative energy one, heralding a post-oil age, a low-carbon world absolutely convinced of the need to limit greenhouse global warming gases, to wean the world from fossil fuels, if possible, entirely, and backed by intense political and popular determination to stamp out carbon emissions from all energy sources. The core belief, the almost sanctified Word of this world, is that global temperatures must not be allowed to rise more than 2°C above the levels of 1990 by 2050, and that if they do, climate catastrophe will swiftly follow. If possible, the temperature rise must be held even lower at a 1.6°C increase. Almost any means are therefore justified to save us from this engulfing peril. A beckoning vista opens out of widening horizons of skyscraping wind pylons in their thousands, of harnessing the sun, of giant hydroschemes, of 'harvesting' bioenergy resources such as cassava, soya, sugarcane, and palm oil, of new forms of individualized energy supply, of countless new and ingenious technologies for creating energy abundance from every natural source, large and small, and for using it all with hugely increased efficiency and economy.

This is also a world on the march with the strong backing, at least on paper, of most Western governments (including those who want to be the 'greenest ever'), of the United Nations, of the European Union with its ambitious carbon targets and plans, of G7 and G20 political leaders, of the idealist young, of celebrities and scientists, of environmentalists and naturalists, of religious leaders, seers, and prophets. It is the message from the Paris Agreement, encased in hyperbole and hope.

Meanwhile, the billionaire giants of the modern global economy have joined in the combat, pledging $20 billion on potential green energy 'solutions' and an attempt to achieve a 'carbon-neutral world' by 2050.[1]

The Judgement of Paris

Now we have the commitment or the aspiration of almost all the world's leaders to measures which will result in zero, yes, zero carbon dioxide emissions by 2080 or thereabouts. This is the message backed to the hilt by the UN which has come to us from the 196 countries or so gathered at their climate conference in Paris at the end of 2015. It amounts

1. Announcement by Sir Richard Branson in New York on 1 December 2015 as he launched the new Breakthrough Energy Coalition, including Bill Gates, George Soros, Mark Zuckerberg, Mukesh Ambani, and many other business giants.

to nothing less than an outright declaration of war on the whole planetwide system and the infrastructure of coal, oil, and gas production, distribution, and use, and on every energy company and business involved, the system which currently provides more than 80% of all energy on earth. It pitches trillions against trillions. If intentions and official declarations can be turned into enforceable measures, into tax penalties, levies, legal prohibitions, fines, even blockades, it appears to place the fossil fuel world on borrowed time.

This UN story tells of an existential threat to humankind. It is the ceaseless chronicle of melting ice, drowning islands, engulfing storms, boiling temperatures – and all in the quite near future – seen by some as the worst and biggest global threat of all. It comes in a tone of impatience and intolerance of delay, even of triumphalism, as though a world struggle has been already fought and won. It puts ALL fossil fuels, their extraction, their handling, their processing into power and light, their very consumption, 'on the wrong side of history'.

But here is the extraordinary thing. Both these powerful and determined empires believe the future is mostly theirs. Both are confidently backing their views with trillions of dollars of investment.

There is some overlap and some cooperation, even fraternization, along the battle front.

The oil industry and the fossil fuel camp concede that green and renewable energy sources will have a modest slice of future world energy supply. Today's big oil and gas companies want to show that they, too, can be green. They, too, have made forays into renewable (mostly wind and solar) investments, set up commissions, burnished their green credentials as best they can, vowed to do more.[2] Most energy company leaders, although by no means all, agree that man-made greenhouse gases, chiefly CO_2, threaten the planet's climate if their emission is unchecked. But in their hearts they are convinced that only oil and gas and coal do the job of supplying the planet with the energy it needs. Green and renewable energy sources may help, but the biggest load will always be carried by fossil fuels.

The opposing green camp, for its part, concede that the transformation to a fossil-free age will take time, although they also feel it is all taking far too long. They believe time is running out and action is too slow. Both sides acknowledge that when large-scale 'battery' storage of electricity becomes viable – which is not yet here by a long way – it will make power supply from intermittent wind and solar sources more attractive and reliable.

For some, there is also common ground over carbon capture and storage (CCS). The more moderate elements on both sides place some limited hopes in new technology for separating CO_2 from fossil fuels as they are burned. There is a longing here for the silver bullet, the 'holy grail' that will allow fossil fuel-burning use to continue but with the carbon elements somehow piped away and buried in emptied oil and gas reservoirs under the sea. If that sounds complicated and expensive, it is because it is. Energy companies have tried out a few experiments, and governments have spent prodigious amounts of taxpayers'

2. 'We have been too absent from the debate', said Shell's CEO, Ben van Beurden, at the London Oil and Money Conference in October 2015. Similarly, Bob Dudley, CEO of BP, looking forward to the Paris climate conference, said, 'Something substantial needs to be done.' Dudley was speaking after six of the European oil majors had written to the *Financial Times*, calling for 'widespread and effective carbon pricing' to be agreed in Paris. American oil majors did not join in.

money. But no one has so far found a way of bringing costs down to levels that would be remotely commercial or competitive.

The CCS issue – of which much more later – is of course anathema to the more extreme opponents of fossil fuels, since, if ever successful, it would allow the continued burning of coal, oil, and gas for electricity with the CO_2 emissions harmlessly siphoned off.

Some other bridges are being forged by technology, such as a large-scale Shell project in Oman to use solar power, not as a substitute for oil and gas but to help oil production by pumping solar-heated steam into the stickier oil wells – a nice example of collusion, intended or not, rather than collision.

Then there is nuclear power. Both sides uneasily acknowledge some role for nuclear power generation – the post-war seventy-year-old dream of cheap atomic power for all that somehow has never quite matured. Greens fear all kinds of China syndrome-type dangers but accept that nuclear electricity is low-carbon power. Oil and gas leaders fear the titanic costs of building nuclear power stations, but recognize their long-term sustainability.

But if in these areas there is some degree of common outlook, that is where amity ends. It is vastly outweighed by growing worldwide tensions and outright hostility between the two fiefdoms, the two cultures, the two tomorrows.

Will these two worlds collide? Will they fight each other? Or can there be more bridges between them? The overriding need of the developing world is for cheap and reliable energy supplies to lift millions out of poverty. So which will be the cheapest and best energy source to bring that blessing? If it is oil and gas and coal, will the crusade against fossil fuels stand in the way? Or will alternative renewable energy sources at last become cheaper and more accessible? How can a head-on clash – an unnecessary conflict – be avoided between the commitment to reduce world poverty as fast as possible, and the commitment to reduce, indeed phase out altogether, the sources of greenhouse gases? The Paris 'deal' of December 2015 reaffirms, at least on paper, the intention by the world's leaders somehow, by hook or by crook, to prevent the planetary atmosphere warming more than 2°C above the levels of 1990, and to do this by bringing the emissions of carbon from burning all fossil fuels to a virtual halt, if not by 2050, then soon after. As we shall see, their chances of actually doing so are vanishingly small.

Indeed, on all present trends, they are bound to fail. This is because however draconian the measures against fossil fuel-burning, carbon emissions from a thousand other sources connected with everyday existence – in fact, from the very act of breathing – will continue to grow. Indeed, 30% of all carbon dioxide emissions come not from fossil fuel burning at all but from food and agriculture. And there is no possible or conceivable way in which a growing world population's need for and thirst for electric power can be checked.

So we are now heading fast for a war to be waged which neither side can possibly win.

Is there a more balanced way through to a new energy pattern? These are the questions this book will explore, and reach some surprising conclusions.

Along the way, we will meet some astonishing paradoxes, contradictions, and conundrums.

We will discover that despite all the dire warnings about energy shortage and 'peak oil', despite the ever-rising costs and risks, despite billions of impoverished people still lacking basic power supplies (with the forecast that even by 2030, at least a billion will still lack any electricity supply, 600 million of them in Sub-Saharan Africa), despite the outright conflicts round the world over energy sources, there is, in fact, an abundance of energy, *no shortage at all.*

We will discover that the astounding and life-changing information revolution and the digital technologies, as they race ahead, are having the same fragmenting and centrifugal effect on the huge energy monopolies and their markets as they are on other industries, from banking to hotels, from retailing to transport, from mass, one-size-fits-all production methods to consumer-driven approaches. Once integrated and global, energy markets are being broken up and balkanized. This could also be the fate over the next decade of the giant international energy corporations that used to dominate world energy, if it is remembered that, as in most other fields, it is not the strongest and fittest that survive but the most adaptable.

We will visit the impossible dilemmas of energy policymakers – in particular, the so-called trilemma (or triple dilemma) which seeks at one and the same time to provide affordable and reliable energy supplies, and yet secure low carbon emissions from all power generation – and the crashing failure to resolve this trilemma in many countries, Britain and Germany in particular, as they wallow in their own special policy quagmires.

It will be shown how the resulting policy ambivalence and conflict are having perverse and damaging results, with peoples, jobs, and the environment all paying a very heavy price in consequence.

At the same time, deeply disorienting issues are cutting across the energy scene, such as the shale oil and gas revolution in the US and the consequent world oil price collapse, such as growing Russian aggression, terrorist dangers, nuclear power doubts, Middle Eastern chaos, the rise of China and India, the huge costs of going green, and the hectic and unending advance of new technology.

Energy Empires in Collision describes how both the European Union and several of its member states, including Britain especially, have become entrapped in costly policy contradictions and muddles.

Above all, we will uncover some of the myths, the half-truths, and the outright deceits that permeate all sides of the energy scene. How rigged is the global oil market? Is OPEC really in decline or is it playing a long game? Why is energy so expensive if it is so abundant? Is there really a global transformation taking place? Why have carbon emissions and other greenhouse gases, far from falling, actually risen by 60% since 1990? Why are they still predicted, however hard the efforts to curb them, to rise well above the safety level about which armies of scientists warn by 2050? Why have events again and again upset 'expert' forecasts? Why is the use of coal growing so strongly and so fast when coal –the black diamond – is so feared and unloved and meant to be shrinking?

Why do well-intentioned government interventions again and again make things worse and invariably create unintended consequences? And are green policies really green? What do Western energy policies contribute to reducing carbon-modest emissions and at what true cost? Is Asia, rather than the West, the region where the future of global energy is really being decided?

Policy Perversity

The answers – where they exist – lie not in basic energy resources but in confused energy policies and politics, in ambiguity of aims, in trends and facts suppressed or ignored, in repeatedly bungled reaction to events – often unanticipated and which may be far outside the energy scene – and in deeply mishandled environmental developments. What too many of those intervening in different aspects of the whole energy nexus often miss, however well meaning their intentions, is the sheer interwoven complexity and fragility of the whole energy scene. If there is one lesson, above all, that the past decade teaches, or ought to have taught, policymakers and governments, it is that in the world of energy the great law of unintended consequences is very, very powerful.

The price of oil, as we shall discuss more deeply, is, of course, very central to the story. It shapes the market, guides the investor, and determines the way the energy world is going – and much more besides. Ultimately, it is decided by supply and demand. But the question is – who or what determines supply? And who or what determines demand?

Once governments step in on either side of the equation, taxing, subsidizing, guaranteeing, seizing, controlling, protecting, prohibiting, promising, disputing, mixing with other priorities – once these things happen – energy issues spiral into complexity, danger, and often chaos. Politics and the state replace genuine prices and the market. Perversity is the most likely outcome. Unintended consequences abound.

The recent past is littered with examples. Attempted price freezes on fuel result in higher prices. Official moves to cut carbon emissions lead perversely to higher emissions. Green policies lead to brown and black results. Subsidies for 'clean' fuels result in more 'dirty' – i.e. carbon-intensive – fuels being burnt. Government interventions to bolster security and reliability lead instead to less security and more power cuts and shortages, crippling the progress and lives of whole communities. Attempted 'help' by the state in energy markets leads to hindrance and damage; measures to increase fairness make life more unfair and so on.

Worse still, political interventions confuse and disorientate commercial decisions. The pattern of investment needed both to extract primary fossil energy from the earth and harness power from natural forces becomes disrupted. The vast sums required to be committed, usually over exceptionally long periods, to turn energy into power and reliable electricity flow, fail to materialize.

Much of this is inevitable, although worse than it need be. Governments are bound to be drawn right in. Agonizing decisions with immensely long-term consequences, far outside the range or rhythm of democracy, the shorter-term perspective of public interest, or the upheavals and turmoil of passing events, have to be somehow agreed. Energy resources of whatever kind have to be developed, financed, processed, transported, transmitted, distributed. Colossal and very long-term investments have to be made and projects, embarked upon. Primary energy – gas, coal, wind, sun, water power, uranium – has to be transformed into secondary energy, into electricity – and electricity becomes life. Whole societies depend on the absolute, guaranteed security of supply, and in the modern world, whole societies have come to depend on the security of demand.

A Warning Bell

Half the world now lives in cities. Electricity is its oxygen. 1.2 billion more people lack it and long for it. For city dwellers, any interruption is chaos. Power interruptions bring cities to a standstill, deeply damage all civilized communities – and increasingly primitive ones as well. There is no corner of modern living which does not depend almost totally on reliable electricity supply, and no corner of the earth which does not need it. No wonder energy is politics – politics, national security, money, and human progress. And no wonder governments find themselves at the epicentre of energy matters.

Cutting across all this, the political message about a greener world has become hopelessly confused. In Europe, energy costs have soared and the dangers and disruption of primary supplies and interruptions in power supplies have multiplied – in sharp contrast to the American scene where energy costs have plummeted. American gas prices are now at their lowest in modern times –although an intense debate about exporting cheap US gas might alter this picture.

In Britain, the mishandling of both energy and climate issues has produced confused and worrying results, threatening the reliability of electricity supplies, actually leading to more coal and diesel oil being used for power supplies, intensifying social injustice, weakening competition, and destroying livelihoods, and all without much of a contribution to the environment. In fact, environmental and countryside desecration has become the hallmark of the so-called green transformation, emphasizing the vast complexity of energy matters and the maze of good intentions paving the road to bad or opposite results.

Despite world energy plenty, the policy labyrinth has become deeper and darker, the right paths to turn to and the way out, still more difficult to find. In large parts of the developing world, easy access to cheap power remains a dream with the numbers of those relying on traditional biomass fuels for cooking – often wood and dung collected by hand and often with toxic consequences – still at 2.8 billion and actually rising as the result of high cost constraints on fossil fuels. The green policies in one part of the planet have turned into the brown and backward consequences in another part.

Is it all worth it? The question hangs in the air. As the costs of energy 'transformation' rise far beyond earlier optimistic estimates, are they still just a modest 'insurance premium' to safeguard the rights of those unborn, and protect our planet from destruction, as claimed in the famous Stern Review?[3] Or an unnecessary and unfair burden on today's poorest and those least able to bear it?

Upsetting the Forecasts

Aside from these serious policy doubts, the whole energy landscape is in upheaval. The onward march of technology has opened up vast new sources of commercial supply of energy in many different forms, whilst at the same time reducing the growth of world energy demand. Future projections based on the past have been once again falsified. The financial crisis and economic slowdown of the past few years have changed the world energy balance, with flat demand in the so-called advanced economies and oil prices first soaring (at one point to $140 a barrel) then collapsing (dipping to $30 in 2008 and again below $40 in 2014 and 2015)[4] and sending investment plans and projects scurrying back into cold storage.

Meanwhile, the shale gas and oil revolutions (using improved technologies for fracturing deep shale rock formations) have altered global supply patterns almost beyond recognition. Not only has the USA become the biggest producer; no less than forty-one countries are now developing their shale gas and oil potential, including countries in Pacific- and Central Asia, in Africa, in Latin America. Australia has moved yet further ahead as a major energy power, supplier-in-chief of coal to China.

3. Sir Nicholas (now Lord) Stern, *The Stern Review on the Economics of Climate Change*, October 2006.

4. With some divergence either side of the Atlantic between Europe's benchmark, Brent crude, and the usually slightly cheaper American West Texas Intermediate. The gap varies.

American oil import needs have shrivelled, with the prospect of self-sufficiency in gas and oil and switching to exports now having a definite date not far ahead. This alone has already changed not only world energy parameters but also contributed decisively to geopolitical shifts in power and influence.

Rapid falls in world crude oil prices, with knock-on effects across the whole energy supply system, have sent many financial calculations and project plans into a spin, shaken governments, altered trade flows, and tilted once again the international distribution of power and influence. Havoc has been wreaked on whole energy provinces, such as the North Sea and Canada's oil sands industry. More than that, the once proverbial power of OPEC to shape world oil prices and act as 'swing' producer of oil has been called into question, ushering in a new era of volatility in terms of oil prices and hence in terms of all energy prices and plans.

The puzzling thing is that oil prices have continued to fall even while supply sources are constrained in many areas by political turmoil, environmental resistance, and outright conflict. Iraq's ambitions for becoming one of the big three world oil exporters have faded. Iran, with some of the world's biggest oil and gas reserves, has been tied down by sanctions, although it is now being gradually unshackled. Egypt has almost stopped producing hydrocarbons and is importing gas from Israel, although this may all turn around as big gas field finds off Egypt's coast are opened up in a year or two.

Canada finds problems in shipping out its oil from its Alberta oil sands operation, and the economics of their oil sands operations have anyway been shattered by the oil price drop. Libya is bisected and output has been halved. Nigeria has cut output as it struggles with violence and terror from the north. The Russian economy, as we shall see, has been severely damaged. Venezuela has been bankrupted, with shortages of even food and basic materials.

If even a few of the current obstacles to oil and gas production become unblocked, world supply will swell even further, while demand could be even more restrained. For example, if Japan moves forward to restoring the bulk of its nuclear power supplies and reopening some of its forty-three operable nuclear power plants, as it is now tentatively starting to do, its current huge thirst for imported gas and oil will fall sharply – and world demand will fall even further. These trends point ever more clearly towards a prolonged era of low fossil fuel prices.

Only so-called 'high-impact' events – terrorist violence striking at key installations or transmission routes at the heart of the international oil system or whole nations collapsing into chaos – might act as a brake, reverse the slide, or produce a sudden and temporary oil price 'spike'.

There was a time when a simple production cut by the Saudis would have stopped the price decline. Now it makes no difference: others simply step in and take their market share. So it will not happen. Instead, the leaders of the great Arab oil-producing states now wait hopefully for the other non-OPEC producers – by which they mean primarily the American shale producers of both oil and gas, and the Russians – to 'act responsibly'.

Some say this is their considered strategy: to throttle high-cost American shale oil and gas production as well as green rival sources, and then reassert the place of cheap-to-produce oil in world energy – but at a resumed high price. But will that day ever come back?

The poorer and higher-cost oil-producing states, such as Mexico, Venezuela, and even giant producers like Russia and Nigeria, relying most heavily on oil revenues to balance budgets and keep governing elites in power, have felt the ground shifting beneath their feet. What are they going to do? Cut output or just pump and sell more and more at lower prices, in an effort to keep the revenues flowing? We will return several times to this absolutely central question.

Over on the green side, with times of austerity closing in, governments everywhere have found it infinitely harder to impose green costs and levies or to find finance for the great energy transformation so ambitiously planned earlier. As oil and gas prices have fallen, the sizes of subsidies for alternative lower-carbon energy supplies have grown larger, generating rising discontent from consumers who have to pay up. Perhaps the worst and most counterproductive aspect of this has been the high-handed assertion of 'experts' that the whole, gigantic transformation of the world energy pattern needed to combat global warming could be achieved quite cheaply and painlessly. A good example is the statement from the British Government's Committee on Climate Change that emissions can be reduced 'at relatively low cost'. Another recent example comes from the eminent economist Jeffrey Sachs who asserts that when it comes to reducing CO_2 emissions below danger levels by 2050 or 2070, 'the cost will be modest'.[5] Numerous other 'expert' voices have insisted on the same view – all, it turns out, utterly, utterly misleading and guaranteed to provoke a backlash of protest from disappointed energy users everywhere, as has duly occurred.

Such views seem to have been propagated in a kind of cocoon, utterly detached from the reality of real-life industrial energy costs, home energy bills, and the desperate needs for cheap power which a fifth of the world confronts, as will be explained

5. Jeffrey Sachs, *Financial Times*, 30 November 2015.

in detail in later pages. Blithely ignoring the true cost factor in global decarbonization places a falsity at the centre of the process and seriously undermines both the green case and the green cause.

A second argument from this quarter is that technologies are just around the corner which will bring down costs dramatically. But a more candid realism would have to admit that there are many corners ahead and not all of them will be turned.

Meanwhile, big shifts have taken place in the geopolitical energy pattern of the past. Not least of these is that the position in the Middle East, for a century past the cradle of oil supplies, with two-thirds of world reserves, has been challenged. Repeated and growing Middle Eastern instability has raised new questions about supply lines. Iraq has seen a third of its territory overrun by Islamic extremists and is still riddled with unending violence. Syria is still in flames, with no end in sight to an increasingly complicated and many-sided conflagration, which has left the powers floundering and impotent. Libya and Yemen remain deeply unstable. Egypt may have gone through the full revolutionary cycle, from tyrant overthrown to liberty hailed on the streets, to anarchy, and then back to a new strongman picking up 'power from the gutter'.

Above all, the most ruthless brand of terrorism, intertwined with Islamic religious schism and hatred of the West, bursts out from the Islamic State (Daesh) strongholds to inflict merciless death and disruption, spreading its franchise far beyond Iraq and Syria to Libya, to Algeria, to Kenya, to northern Nigeria, to Turkish towns, and to Tunisian beaches – as well as to Paris and the heart of Europe.

Elsewhere, the Daichi Fukushima nuclear power station disaster in Japan has added additional uncertainties and costs to civil nuclear developments round the world (see Chapter

Eleven), while coal-burning has leapt not only in China and India, where hundreds of new plants are being planned, but also in Europe (although not in the USA, which ironically has all along refused to subscribe to legally binding carbon reduction targets!).

Until recently, discussions of global energy tended to be focused on the oil needs of the Western world. But that, too, has changed fundamentally. It is in Asia that the shape of things to come will be largely determined. The demand for oil (and gas and coal as well) will rest on the appetites and development paths of Asian nations and their citizens, lifting themselves out of poverty at last, entering the fully industrialized age, and even vaulting over it into new technologies and new patterns of production and dominant positions in world markets.

The growth of carbon emissions, and the impact on global warming will likewise be decided by the pattern of Asian industrialization and the rise of Asian consumer societies. Whatever deal governments such as China and India have entered into on paper at the recent UN-led Paris climate conference, forces beyond the control of governments may decide matters. The pattern of progress may be similar to the Western path, or it may be utterly different, far less energy intensive, avoiding altogether what the author Naomi Klein calls the West's 'Dickensian phase', its era of dark satanic mills.[6]

Scarcely noticed in the West, new battlegrounds are forming on which the world powers are testing each other over control of oil and gas resources. Russia is desperate to diversify its gas exports away from Europe and into Asian markets. But it has rivals and competitors there too. In such places as Kazakhstan

6. Naomi Klein, *This Changes Everything: Capitalism vs. the Climate* (Penguin, 2012).

and Turkmenistan, Chinese, Russian, and American players jostle for influence. For these regions, as for many other countries round the world, either dependent as they are on oil and gas exports or living on the hopes of oil and gas revenues tomorrow, the news about the other future, a future without fossil fuels and with new entirely green energy sources, where people want far less oil, has simply not arrived or been noticed, although when and if it does, the impact will be immense.

The time has clearly come to work out new ways of bringing some stability back to a chaotic energy landscape. Those ways will be hard to find. Events, compounded by devastating policy and political error, conflicting aims, and misunderstanding have bedevilled the scene. Animosity between the two energy worlds – the established fossil fuel systems and the green, low-carbon challenge – has intensified. But there could still be a collaborative pathway forward which this book seeks to illuminate. That is the aim here – no lobbying for any particular case or instant solution, no false claims, just a patient untangling of some of the misguided beliefs, proposals, and policies which still lead us astray and divide us, and from which we must now move on or instead face truly frightful penalties.

Lord Stern asks in his recent book: 'Why are we waiting?'[7] If climate change threatens us all, why is the world delaying in closing down the fossil fuel age, halting the growth of carbon emissions, and giving the world some chance of keeping within the 2°C rise by 2050 beyond which appalling climate chaos is predicted? Why the inaction?

The answer is that nations, peoples, businesses, families, ordinary folk are waiting for guidance and policies which lift

7. Nicholas Stern, *Why Are We Waiting? The Logic, Urgency, and Promise of Tackling Climate Change* (MIT Press, 2015).

rather than crush, which enhance prospects rather than narrow them, which raise living standards rather than freeze or lower them, which reconcile sensible climate policies with sensible energy policies, rather than putting them in head-on collision.

The green cause is powerful in ruling circles and in richer countries (and paradoxically amongst student activists) but its methods are not popular. People like its vision and idealism but they do not want to pay for it or suffer for it. In the meantime, the world empire of fossil fuels – oil and gas and coal – will continue to meet the changing world's needs for energy to fuel economic growth. This will be done with efficiency, immense skill, and experience, with the latest technologies and quite probably at decreasing final cost to energy consumers.

If the greens are to compete and make a real impact on climate change, they, too, must do more than call world gatherings, such as the Paris climate conference event, and pass resolutions. They will need to cut costs and bring ever more ingenious technologies into play which produce immediate and tangible benefits for all, and which add to wealth rather than subtract from living standards, which help rather than hinder the poorest, which protect and respect all aspects of the environment rather than damage it, which help rather than penalize an increasingly urbanized population.

They must ally with, rather than attack, the existing energy industries, drawing on their same skills and innovative momentum. As the pages ahead repeatedly emphasize, they must work *with,* rather than against, the natural capitalist market process. The planet cannot afford schism. Neither side can crush the other. There will be no complete fossil fuel elimination, or anything like it, as many green campaigners, including governments, demand, and, at the other extreme, no green marginalization, as some oilmen, gas producers, and coal kings complacently seem to think. For a sustainable and balanced future which reconciles the world's commitment to reduce poverty with the green transformation of energy, the two empires must unite, not divide and collide. Then the waiting will be over.

A Summary of the Chapters Ahead

Chapter **Two** begins in the Middle East and the Gulf States, seen until recently as the cradle of world energy supply and the trigger region for numerous world energy alarms, the kingdoms and sheikdoms sitting on by far the biggest oil reserves and with the easiest and cheapest oil access. How will this whole region fare when faced with the triple challenges of encroaching and paralysing political instability and violence, swelling non-OPEC oil and gas production (especially from American shale oil and gas), and world opinion and governments turning generally against fossil fuels, as marked by the Paris Agreement of December 2015? Are low oil prices a deliberate strategy for the long-term re-enthronement of Middle Eastern oil and the crushing of rivals, or part of an unavoidable and different future?

Chapter **Three** further describes the wider upheavals that have changed the face of energy, its sources and supply, its distribution and its use – all in very recent times – and altered significantly the challenges facing policymakers and decision-makers. The true coal story, hidden or ignored in the global warming debate, is uncovered.

Chapters **Four, Five,** and **Six** address the triple 'horns' of the so-called 'trilemma' (double-layered dilemma) facing energy and climate policymakers worldwide, and their continuing and utter failure to resolve it. This declared 'trilemma' is how to combine three apparently conflicting aims of energy policy – affordability, security and reliability of supply, and lower-carbon emissions. Why has progress been thwarted on all three fronts? What is the fate of the poor, the deprived, the frightened, the threatened – in short, about a quarter of the world's population still living in abject poverty, struggling daily to survive, always under threat from sickness, disease, food and water failure? Are these the real victims of the trilemma policy failures? There used to be a time when those on the Left of politics, with the readier compassion they claim for themselves, could be counted on to uphold the cause of the poorest, the underclasses, and the dispossessed. But not any more, it seems.

Chapter **Seven** brings us to the European energy imbroglio. False starts, contradictory aims, and unintended results abound. We analyse the ambivalent tangle into which EU energy policies have become entwined, with energy cost penalties on business and enterprise which the former EU Industry Commissioner, Antonio Tajani, has described as 'a massacre'. We look at the European Commission's latest effulgence, the European Energy Union, described by one member of the European Parliament (by no means a Eurosceptic) as 'a ragbag, of measures administered by schizophrenics'. The chapter examines so-called carbon leakage (the transfer of heavy carbon dioxide-emitting industries to the developing world) and European attempts to 'lead the way' in carbon reduction through emission trading and carbon taxes – which instead have led to economic damage, painful competitive disadvantage, and higher coal-burning, while doing very little to cut carbon.

Chapter **Eight** goes further inside the European disorder and looks in particular at the Russian shadow over European gas supplies (asking what Vladimir Putin and his circle really want for Russia), at the muddle in a Germany which has denied itself nuclear power and has found a truly horrifying 'answer' in burning more coal – of the dirtiest kind – than ever in recent times, and the Polish frustration with past and current EU energy strategy, now being transformed into major political change in Poland.

Chapter **Nine** comes specifically to the British energy scene which has been described as 'stumbling into crisis'. It explains what has gone so wrong and how precarious the British power situation has been allowed to become, how government plans for the revival of nuclear power are faltering, and how its complicated plans for maintaining power supply security, balancing the whole grid system, and encouraging more power investment, are, or are not, working out.

Chapter **Ten** outlines how some sort of rescue plan for European energy policy can be pieced together – as part of the overall rescue and reform which the whole European project now urgently requires.

Chapter **Eleven** shows how the price of crude oil still dominates every aspect of the energy scene, shaping investment, shaping consumption patterns, shaping the global economy. It examines how the price is formed by a bewildering and changing brew of economic, political, and random factors. It surveys the world of the new producers and the new frontiers in energy supply – already greatly altered from the pattern of previous decades – and shows how the current halving in the crude oil price sends powerful waves through all the world's economies, and it faces the question to which no one really knows the answer – where does the oil price go next?

Chapter **Twelve** goes to the heart of the green campaign's prospects. Politically, the case for a global transformation of the world's energy is winning. But in practice, can the costs of renewable energy, such as solar and wind power, ever be brought down to levels competitive with coal and gas prices? Can coal and other fossil fuels continue to be burned after all, but with the carbon all neatly captured and siphoned away (the CCS question)? We turn to some 'renewables truths' and some of the realities – both negative and positive – behind the claims and counterclaims for wind power, solar power, geothermal power, biomass, micro-generation, and all the other non-fossil fuel sources, not least the biggest of all, hydroelectric power. The latest prospects are evaluated for clean coal-burning, carbon capture and storage, electricity battery storage, and other frontier technologies.

The central question stands out. Can renewable and weather-dependent energy compete in the market without subsidy and unbearably large governmental support, with its inherent unreliability as ministries and moods change? And might not the biggest step to lower carbon come not from the supply side at all, or from green energy sources, but simply from a quantum leap in fuel efficiency and economy?

Chapter **Thirteen** deals with the nuclear conundrum, or biting the atom, as James Lovelock terms it. Green opinion is divided as to whether low-carbon nuclear power should be lumped together with other low-carbon sources. Strong and sometimes irrational fear of nuclear power and its associated waste handling makes many green campaigners want to keep well away from nuclear. But the awkward realization that carbon reduction targets are never going to be reached without a nuclear component pulls the other way. That tears green campaigners and low-carbon policymakers in two directions.

The chapter, therefore, traces the Fukushima saga and the prospects for civil nuclear power – in fact, so far only partially checked globally by the Japanese disaster. In some countries, such as Germany and Austria, the stop is complete; in others, there has been no disruption, with new nuclear projects being pushed ahead faster than for many years past. It is also the story of nuclear power innovation, new designs, and construction techniques from Korea, Japan, China, and Russia, and their impact on the West, with China, in particular, pressing hard to get into Western (especially British) markets. It questions whether the giant station planned by the British, together with EDF at Hinkley Point (last of the dinosaurs?), will see the light of day, and if it does, whether the Chinese will come to regret their involvement. Or has nuclear power got a different future?

Chapter **Fourteen** brings us to the Chinese scene. It shows how all energy roads used to lead to the Middle East but now lead to China, and it follows its worldwide grab for energy resources, as well as its unfolding role in opening out the whole of East Asia. Is it here where both our energy future and our climate future will be finally shaped? Can China keep to its commitment to check the growth of carbon emissions, of which it is now the greatest source? Are the figures to be believed?

Chapter **Fifteen** ('The Tomorrow We Deserve') – sets out the book's recommendations and maybe hopeful conclusions. But 'conclusions' is a risky word to use in an area where change is happening so fast, where technology is opening up new avenues so rapidly, and where new implications and outcomes spring up so widely that no concluding point is ever reached.

Nevertheless, governments, west and east, north and south, and their policy inspirers, can at least do some things – and stop doing others – to help progress, clear up the present chaos, and make better outcomes more likely. Not everything by any means is within their power. Indeed, power has been slipping away from authorities and institutions in many areas, energy developments included. But governmental and policy power also remains, and the message on these fronts is clear.

First and foremost, admissions of error are the beginning of wisdom. Vast and inefficient subsidies on green energy sources have achieved little, and far smaller sums spent on research into cheaper alternatives would yield much better results. However fierce the measures which governments may pledge themselves to take in order to curb carbon emissions, they are not going to cover everything. Although that is not what governments are admitting, carbon will go on pouring out from other sources into the atmosphere. Demonizing all fossil fuel-burning and forcing consumers to pay the earth for still ultra-expensive renewable energy turns out to be NOT the right way to meet the low-carbon challenge.

Secondly, the curbing of impatient and utterly counterproductive green zealotry is essential. Thirdly, new technologies and global market forces will do more, if allowed, to drive energy transformation than a dozen bundles of interventions, regulations, penalties, and official pressures.

Above all, Chapter **Fifteen** returns to our main plea – that the immense empires of hydrocarbons and of green world ambitions and hopes should work together rather than fight. Cooperation is in the direct interest of the vulnerable, the poor of today, and the struggling and challenged of tomorrow.

Finally, Chapter **Fifteen** also takes a hesitant and limited look forward.

Forecasting the future shape of energy supply and usage is a fool's game but we allow ourselves a limited and not entirely unrealistic look a mere ten years ahead despite the great uncertainties and obvious dangers and despite the need for many big energy projects to plan over decades.

Peering ahead through the mist of future surprises, of unending human ingenuity and constant innovation, we outline the likely and attainable framework for an unworried world of energy plenty for all.

Some hope, admittedly. Did the experts, the heavily resourced think tanks, the academic oracles, the government advisers, the armies of opinion-formers even foresee the internet revolution, the financial collapse, the Arab and Middle Eastern chaos, the oil price vertical drop of 2014-15?

But hope springs eternal and it is just possible and useful to see, at least a few years ahead, where we might be going, human folly allowing.

CHAPTER TWO

The Cradle Rocked

The Middle Eastern oil kingdoms in crisis – the triple threat to oil and gas riches – encroaching regional violence and instability – swelling non-OPEC oil and gas production – opinion and governments turn against fossil fuels – the old assumptions of Middle Eastern energy dominance fade – the post-2008 shocks – oil and gas demand weakens – the impact of Asian growth and China – has OPEC lost control? – the clever Emiratis who ride both horses.

Sheikh Ahmed Zaki Yamani, the Saudi Arabian oil minister, spoke in a low voice. He had something important to tell me, something I ought to know – now, this very moment.

It is January 9th 1980. We are travelling in his car towards the airport. We are coming to the end of an official visit to the kingdom. Sheikh Yamani has been a gracious and assiduous host to me, the UK Secretary of State for Energy, these last few days, coming to meet me in his private plane at Dhahran in the north, where my party and I arrived, taking me to see the king (King Khalid, fifth son of the founding monarch of the House of Saud, Abdulaziz ibn Saud), and looking after my every need.

Now it is time for my mid-morning departure. Yamani leans forward a little beside me and adjusts his keffiyeh. You should know, he says, that just now in the last hour or so, most of the prisoners captured after the Mecca mosque outrage last November have been beheaded – sixty-three of them to be precise – all publicly and in various cities round the kingdom.

Suddenly, the morning did not seem so bright. I was being given a momentary glimpse into a world of fanaticism, violence, and revenge – a primitive world that seemed aeons away from oil production, refinery technologies, price and demand forecasts. An eye for an eye, a tooth for a tooth, and no mercy for those who might upset the delicate pact between the ruling family and the Wahhabi clerics in all their Sunni zealotry underpinning the Saudi state. So, swift death to the blasphemers, the false prophets, the profane desecrators (in this case, the ultra-extreme sect who had proclaimed their own Mahdi, seized the Grand Mosque in Mecca, Masjid Al-Haram, the autumn before, killed dozens of pilgrims and soldiers, and held out for two weeks before being gassed out with the help of French anti-terrorist experts).

I hoped there would be no more as no doubt the Saudi authorities hoped. But we were quite wrong. This was one stage in the unfolding decades ahead of bloodshed, religious strife and schism, atrocities on a mediaeval scale, all still with us. And through it all ran and still runs the black river of oil, propelling many of the Arab peoples to unimaginable heights of wealth and power, but also to internecine and inter-religious conflicts of unparalleled ferocity and bitterness.

Thirty-four years pass from that meeting. It is now October 2014. We are again on the way to see the Saudi ruling house – this time in the shape of Prince Salman, governor of Riyadh, twenty-fifth son of Abdulaziz, and rumoured, although no one could be sure, to be the likely successor to Abdullah (the tenth son), now at a great age and in very poor health. Soon, the rumour will become fact and he will be the next king.

In my pocket I have a cutting from the Riyadh English language morning newspaper – an item of utmost significance for the Saudi rulers and their nation and for the whole global energy story. The International Energy Agency in Paris reports that American oil and gas liquids output reached 11.8 million barrels a day in August 2014. In October it would be, on average, 12 million barrels a day. The latest Saudi figure is 11.5 million barrels. Taking crude oil alone, aside from all the other petroleum liquids, such as ethane, the figure is, for America, 8.8 million barrels of crude per day; for Saudi Arabia, it is 8.5 million barrels a day. The US has overtaken Saudi Arabia. There will be consequences.

At the governor's palace, we pass through door after door, room after room, each heavily guarded, each of which opens to another courtyard, until we reach the final waiting area. Courtiers hover. Voices are low. TV men, lugging oddly dated tripods and camera equipment, wait deferentially. Someone whispers. It is time to go in and meet the governor and king-to-be.

This is wealth beyond imagination, beyond the dreams of Croesus, the wildest fantasies of the Arabian Nights, the realm of numbers, the grasp of the human mind. Decades of vast oil revenues have swept lakes of funds, oceans of cash into the hands of the kings, princes, and emirs of Arabia – and most of all, into the coffers of the sons and grandsons of the great Ibn Saud, head of the House of Saud.

This ought to be the inner court of happiness, the throne room next to heaven where all is peace and plenty.

Yet there is unease – a sense of dangers and disturbances to come. Gone is the mood of confident supremacy of the Yamani days. The governor is a younger brother of the king, yet he, too, seems weary. No doubt he senses the tribulations ahead. He keeps reaching for a Kleenex. But that is nothing new. Tired, old rulers have come and gone, but the wealth cascade has always continued, always grown bigger.

But something is challenged. The 'always' no longer seems so certain. Faint tremors from the shifting foundations of the entire glittering structure of power and money run through the courtly scene. A seemingly permanent world of unlimited thirst for unlimited amounts of Middle Eastern oil at ever climbing prices is just beginning to rub up against new realities – the realities that the days of Arabian oil dominance may be passing, that prices may be set by larger forces, and that the world's energy demands may be met in other ways by other regions from other technologies and sources. Middle Eastern oil may be fighting its last battle against both oil and gas from elsewhere and from the green juggernaut.

And beyond energy issues, the politics of the region all around the kingdom seem to be turning sourer by the day. The old enemy, Iran, and the old friend, America, have been negotiating, making a deal. The schism of Islam between Sunni 'south' and Shiite 'north' is everywhere being accentuated. Yemen, right next door, is in flames with the Saudis now engaged in an outright battle for control with the Iran-backed Houthis. Right at home, within the kingdom, the high Sunni dogma of the controlling Wahhabis, the sporadic violence, and the bombed mosques in Shiite-minority areas are all constant reminders of the enemy within.

Further north in the Arab world, the patterns of friend and foe are even more confused. President Bashar Al-Assad of Syria – the part-client of Iran as well as the protégé of Russia – is clearly the enemy so his attackers should be the friends and allies. But his attackers include the barbaric murderers and crucifiers, Islamic State (so-called) or ISIL or Daesh. They may be Sunni-inclined but they are just as hostile to the Saudi state as Iran. So are they friend or deadly foe?

It's the same situation in Iraq where the Iran-friendly and Shia-inclined Government in Baghdad is under assault from ISIL. Whose side should Saudi Arabia be on? Who do they back? From where does the greatest threat come?

A bow and handshake, a swirl of *dishdashes* and *shemaghs*, a volley of camera flashes, a few murmured words about world affairs, tiny cups of coffee poured with a flourish, tipped and taken, hands usher towards the double doors. It is over.

The questions hanging over everything are different from those in the earlier days of Yamani. Then it was how would power be used in a ruthless kingdom, successive oil shocks having cowed the oil-consuming West into paying ten times, twenty times the previous price for the Saudi black gold.

Now, decades later, there were quite different uncertainties. How long could it all last? How could calm stability in the kingdom be maintained with revolution all around, rulers overthrown, civil wars, intensifying religious extremism (not least within Saudi Arabia itself)? What would life be like without the US needing Saudi oil? Would the kingdom's new Asian customers, now taking 70% of its exports, be so supportive? It was new shocks for old, and shocks far less comforting for the hereditary rulers and their families. Where would the old enemy Iran strike next?

The Post-2008 'Shocks'

In the decade leading up to the financial crisis of September 2008, which engulfed almost the whole planet, immense changes undoubtedly took place in the global energy scene. From the mid-1990s, climate concerns and low-carbon aspirations began to transform energy policy; energy companies hastened to strengthen their green credentials; new threats to energy security developed in many areas with the New York 9/11 horror rocking the world. The Caspian and Central Asian region opened up; national oil companies extended their reach; Arctic region possibilities were explored; China's and the whole of Asia's thirst for fossil fuels swelled dramatically. New technologies at every stage of the energy supply, transmission, distribution, and consumption chain were opened up; revolutionary advances in energy efficiency were developed.

These seemed like big changes in the energy world. But they were nothing compared with what was to follow. Radical though these earlier developments seemed, they have been overshadowed by even greater upheavals in world energy in the years since.

First, after 2008 came the fall-out itself from the global financial spasm, in terms of stunted world growth, deep and prolonged recession in the West, soaring debt and stagnation in Europe as the Eurozone troubles shrivelled business confidence. All of these events checked energy demand growth, deterred new energy investment, and almost everywhere weakened government resolve to tackle energy transformation. Oil prices dropped sharply after the financial crisis, triggering cuts in OPEC production, and it was then followed by a stabilization of prices, although at a high level of $100 or so for a barrel of crude. Debt-ridden governments cut back heavily on support and subsidy for renewable energy technologies, leading to accusations of broken faith and widespread bankruptcy in the solar panel industries in several European countries.

Then came Middle Eastern turmoil deepening further and further. Syria began its long descent into total chaos in a many-sided civil war which drags on with repercussions throughout the region – and, indeed, globally with the enormous efflux of refugees, fleeing for their lives and destabilizing not only neighbouring states but the entire European Union.

Egypt veered the same way but was steadied, just in time possibly, by a new military-flavoured regime, but for how long? Iraq found itself partially overrun by the so-called Islamic State (Daesh) with barbarous violence continuing, despite some revival and expansion in oil production. Libya was convulsed, then recovered, and has now been split and convulsed again, with the jihadist extremists bedding in and posing a new challenge to Egypt – and again spewing out a pitiful stream of refugees and migrants as well as jihadis. Yemen has sunk into division and civil war, bombarded by its Saudi neighbour.

The world has watched in horror as the misnamed Arab Spring duly turns to a sickly autumn. Riots and violent protests have spread everywhere from Bahrain to Turkey with Jordan and Lebanon teetering. Streams of refugees pour out in search of safety on a scale unmatched since the end of the Second World War, shaking the borderless ambitions of the European Union to their foundations.

And all the while, the Israel-Palestine issue has continued to pour poison into the regional well, symbolizing the long-drawn-out intractability of every Middle Eastern issue – an ironic contrast to the 1917 hopes of the British Government that setting up a Jewish home in Israel would help cement relations with the Arab nations and improve British standing!

But these deeply troubling events aside, it is on the energy front that the Saudis and all the other big Gulf oil states face their toughest immediate challenge – shale. That was the message burning in my pocket as I talked with Prince (shortly to be King) Salman. The expansion and improvement of hydraulic fracturing techniques and horizontal drilling for gas and oil, leading to soaring American oil and gas output, have altered the pattern of world fossil fuel production – and all the commercial parameters – fundamentally.

Not only is the USA, the biggest global consumer, now set to be self-sufficient in natural gas, and maybe an exporter by 2020, and in oil by 2025 (that is on top of its already massive exports of refined oil products). The new technologies have also vastly opened up or expanded energy production possibilities all round the world – in Africa (Southern, North, East and West), throughout Asia, Europe (including the UK), Australia, Canada, China, Ukraine, the Eastern Mediterranean, Argentina, and many more places. Forty-one countries have now identified both new so-called unconventional shale gas and oil opportunities as well as numerous new conventional sources, made much more accessible by new technologies being applied to oil and gas recovery of all kinds.

It is true that with the price of oil sliding (although for how long is uncertain), exploitation may now be slowed or remain tantalizingly on the edge between commercial recovery and new projects being left as uneconomic. But the overall supply picture is one of massive and continued worldwide growth in hydrocarbons production.

This new energy abundance, combined with forecasts of slowing growth in world demand for oil, has done two things. It has come near to breaking OPEC's monopoly on oil power and, as later chapters explain, it has falsified the basic assumptions on which EU energy and climate objectives were built – which were basically that oil and gas would remain excessively expensive fuels, coming mainly from unstable Middle Eastern suppliers, with renewable energy becoming the early winner. Both assumptions have crumbled.

The hope for the Gulf oil states all along was that new Asian customers would make up for closed American markets and flat European demand for oil. So far China's energy hunger has moved on to a scale which dwarfs the rest of the planet. Huge acquisitions of rights to oil and gas development have been made across the continents, especially in Africa, but also Asia, the Antipodes, Latin America, Canada, Central Europe, and even in the North Sea.

These, went the reasoning, would be the new markets which would keep world oil demand on its upward track.

But here, too, something has gone wrong

First, Chinese demand for oil and gas has not recently been growing as expected. With China's overall growth slowing sharply, oil import demands, while still high (see Chapter Fourteen), have flattened.

Second, Asia has other fuels to burn and other sources to draw on. For example, coal-burning has expanded considerably both in Asia and Europe, with cheap American coal, displaced in the American market by even cheaper shale gas, pouring into Europe and causing the reverse effect – the rapid expansion of coal-burning in place of gas-burning, the diametrical opposite, ironically, of what was intended by climate change campaigners and policymakers. For the first time in human history, coal, oil, and gas will share equally in the primary energy mix, estimated to be each at about 27% by 2035,[1] the rest (19%) coming from renewables (wind, solar, geothermal, hydroelectric, biomass) and nuclear power. India under Narendra Modi has renewed its pledge to produce and burn coal on a colossally increased scale – from 408 million tonnes in 2014 to 908 million tonnes in 2020.

Coal will be the development and growth, the decider. Japan will build more coal plants, arguing that they reduce carbon by being significantly more efficient. Even the United Nations has recognized and encouraged more coal-burning. Coal, as we shall see, is turning out to be the culprit, the back door intruder undermining the whole world battle against climate change. Coal-burning may produce a diminishing proportion of total electric power in the years ahead (it is now about 39%), but the world's consumption of electric power will be vastly greater by 2035, 2040, and 2050, and coal-burning will be vastly greater with it.

As for other suppliers, China can buy ample supplies of all hydrocarbons – coal, gas, and oil – from sources other than the Middle East. Coal and gas come from Australia, although China has recently grown much choosier about the quality of coal it imports from the Australians, but the volumes are still colossal, even though the Chinese economy is plainly slowing down from its previous hectic growth rates. Gas can be piped from Central Asia and now China has struck a favourable deal with Russia for large future gas supplies, Russia being near-desperate to find new Asian gas markets as its European customers turn away.

None of this has deterred the Saudis or their fellow OPEC producers from maintaining high oil output. At 10.5 million barrels a day, Saudi output in May 2015 was almost at a record high.

1. According to BP forecasts to 2035.

But it leaves a corroding atmosphere of doubt. Reflecting this, the kingdom's oil minister, Ali Al-Naimi, told a surprised Paris audience that 'In Saudi Arabia we recognize that eventually, one of these days, we are not going to need fossil fuels. I don't know when, in 2040, 2050, or thereafter'. For this reason, he added, Saudi Arabia plans to become 'a global power in solar and wind energy'.

He might have added that with Saudi domestic appetite for oil and gas growing rapidly (and unsurprisingly, given the heavy price subsidies for Saudi domestic users), there may anyway be precious little left to sell into world markets. Here is one notch up for the ambitions and hopes of the green empire – the nation whose oil-based wealth and power had become proverbial conceding that a post-oil age lay ahead.

And he might have added further that the trajectory of Saudi budget finances was already beginning to look worrying, with a huge (20%) budget deficit looming, projects needing to be cancelled, and assets sold off to make ends meet.[2] 90% of government revenues derive from oil. This is the money 'glue' which counters radicalism, retains loyalties, and buys stability. What happens if it dries up? What happens if the ultracheap domestic price of oil and gas is allowed to rise – a move fraught with political danger but on which the Saudi Government has now taken the first nervous step?

Al-Naimi did not, of course, reflect on a still wider issue no doubt on questioners' lips. Has OPEC, with Saudi Arabia as its kingpin and central player, now lost control of world oil markets? Has an organization which once owned 81% of the world's known (proven) oil reserves, 66% of that in the Middle East, lost the power it once clearly possessed to fix global oil

2. This includes even part of ARAMCO, the giant Saudi state oil company.

prices? Or is there a deliberate strategy in place to drive out all the higher-cost producers round the world who have crowded in to benefit from the recent era of $100 plus per barrel – and thus restore Gulf dominance as the great exporters of low-cost, high-margin hydrocarbons to a still needy world? And if so, is it working?

On the answers to that depend issues of the highest importance not just for Saudi Arabia and the Gulf but for the whole world.

Riding Two Horses

Every year in Abu Dhabi, there takes place a truly massive gathering under the banner name ADIPEC (Abu Dhabi International Petroleum Exhibition and Conference). The conference, which brings together just about everyone involved in the oil industry worldwide, has one sole and central theme – to celebrate the power, potential, and progress of oil, and the mighty cascade of petroleum products which flow from the black, carbon-filled substance that comes from beneath.

'So what?' an observer might say. Is not Abu Dhabi itself the ultimate petro-emirate? Has not oil production lifted a tiny and primitive fishing and pearl-diving community, squatting on the coast, into a gigantic oil-rich power, a world player, a nation courted for its wealth by almost every other, its skyline almost blocked with skyscrapers, its hotels and palaces built on the scale of Ancient Egypt at its peak or the Roman Empire at its most powerful?

Well, yes, but there is a twist to the modern story. Every year, in the same country in the same great city, in the same enormous exhibition centre, there is another gathering of equal magnificence and size, attracting equal crowds of exhibitors, hopeful companies, businessmen, financiers, media folk, ministers, and experts.

But this time the central theme is quite different. It is to celebrate the green energy revolution in all its aspects, the brilliant innovations in countless fields, the new wonders of renewable energy, its production, and its use from every conceivable angle.

Both immense occasions lay claim to the future. At the oil event, speeches assert the unending primacy of hydrocarbons and all their derivatives, drawing on numerous forecasts and expert studies showing that for decades to come oil will remain king, the irreplaceable, endlessly fungible, power source, deploying ever more amazing technologies to recover it from the earth and below the seabed.

At the green event, similar speeches, uttered with great authority, will describe the end of the oil age, the phasing out of hydrocarbons, and the birth of the low-carbon or even zero-carbon world, harnessing the forces of nature to bring green energy to all.

How on earth can both these messages be right? How is it that both messages are trumpeted out from the same place, and with the same organizers behind them and often the same individuals proclaiming them?

The answer, of course, is that the conflict, while it is acted out on the world stage, is in practice completely avoidable and unnecessary. On paper and in wide circles of opinion, indeed, in wide circles of intense and deeply held belief, the two futures may clash. Green campaigners may feel the need to be at war with the whole fossil fuel sector, as the divestment movement shows. But in the middle ground of evolving reality, they can all be made to fit together. Neither side will totally prevail. Those who demand the end of all fossil fuels, a zero-carbon energy world of green power sources, and super energy use efficiency, will never reach their Jerusalem. Those who think the oil and gas revenues will go on climbing in an ever bigger mountain of riches, that money will continue to grow on trees, will find it will not be so.

Those who perceive how the two futures need not be in conflict will work together. The Emiratis with their great gatherings will be able to ride both horses and will do so with skill and dexterity. The oil age and the green age will coexist, although in novel and endlessly inventive new ways, some already visible, some still beyond the shifting horizons of technology.

The clever Emiratis know what they are doing. They are encouraging the application of the most advanced – and cost-cutting – technologies to the fuel oil and petroleum supply chain, from hydrocarbons exploration and recovery right though to industrial and domestic use. And they are at the same time encouraging the application of every innovative technology to renewable energy sources and, again, to their use.

They are indeed backing and riding both horses – and they are right to do so. In time they will deliver, by this dual strategy, more CO_2 reductions, more modest prices, and more reliability than any dogmatic opponent of all fossil fuels could ever achieve. They will provide their own solutions to the baffling trilemma.

The only challenge to this skilled balancing performance is that far outside the circus ring there is a third wild horse roaming, seemingly well outside the discipline or control of the Middle Eastern petroleum states, and indeed outside the whole of OPEC.

Swimming in oil

This is the prospect, daily more imminent, of truly colossal increases in oil production in the Western part of the world, chiefly in the USA. American oil production by 2020 is likely to be around 11.6 million barrels a day, making it level as an oil producer with Saudi Arabia or even ahead of it. The oil price collapse may have shaved some production off this figure, but it remains high, boosted by new levels of productivity.

Growth on this scale could enable the Western hemisphere to meet its own oil needs with ample supply over the next decades, reducing dependence on the Middle East and perpetuating soft oil prices as the norm. But this would tilt the balance in the tug of war between the two empires – oil and gas versus green – back towards cheap hydrocarbons. The pressure on renewable energy sources to cut their costs would be sharply increased.

Far from there being a shortage of primary energy to meet strongly growing world demand over the next fifty years, as bodies like the International Energy Agency in Paris were until recently predicting, this could spell oil and gas surplus, even glut. Unsettling volatility in oil prices could replace the relative calm (with occasional lapses) which the price-supporting cartel of OPEC has imposed over the past four decades. Now its influence could be slipping – at least for a while, and possibly forever.

Together, the Middle Eastern oil-producing states face a triple assault. Regional violence and religious strife surround them; oil oversupply and weak prices threaten them; and green alternatives encroach on them. They have the resources to survive, but do they have the politics, the resilience, and the social cohesion?

CHAPTER THREE

Fresh Upheavals

The power shift to Asia – main energy consumption growth outside the West – climate concerns mount – governments and popular opinion against oil – green subsidies and hydrocarbon penalties – new uncertainties for investors – the 'third earthquake' of hydraulic fracturing and the shale revolution.

The high hopes for a legally binding global agreement on carbon dioxide emissions reduction have remained out of reach. The two-week-long Paris conference on climate change (COP 21 – Conference of the Parties) at the end of 2015 produced generalized assent on the need for further carbon reductions from most countries although not from all, but no global legally binding commitment. While the European Union has tied itself down unilaterally with precise 2050 carbon reduction targets, involving soaring energy costs for the consumer, the world's giant emitters have refused to be so bound and are pursuing their own paths to energy transformation and pollution control.

A climate challenge is certainly recognized by the world's biggest economies – China, India, Russia – and all manner of pledges and ambitions have been duly expressed. But from these great emitters – on whom everything in the way of emissions curbing depends – there has been and will be no precise undertakings. They see different and maybe more effective ways forward. They simply refuse to lash themselves to the same mast of rigid carbon reduction targets as the Europeans have done.

Meanwhile, plans both in Europe and elsewhere for such devices as emissions trading and universal carbon taxation have faltered, as well as complicated schemes for supposedly encouraging lower-carbon activity worldwide. In Australia they have been discarded outright.

The disconnect between green goals and what is actually happening has never been wider. For the first time in modern history, no one primary energy source dominates and no one source of supply dominates. As earlier mentioned, oil, gas, and coal all take equal slices (about 27%), with renewable (green) sources pressing in at the edges. Middle Eastern oil has to share the stage not only with American shale but with new fossil fuel sources, conventional and unconventional, all round the world, as well as with a growing renewable energy sector.

The uncertainties and deep ambiguities of the energy scene have been vividly reflected in the decisions of energy companies to give up trying to make firm predictions and resort to scenarios, alternative path streams, and the like. *New Lens*, Shell's recent massive forward survey, posits two divergent scenarios, depending largely not on actual primary energy flows but on which way the politics will go and how the interaction between government policies to achieve low-carbon ambitions and policies to ensure energy security and contain energy prices unfolds.

The International Energy Agency (IEA) also speaks of a 'blur' in assessing the energy future. It clutches at a 'reference case', which essentially extrapolates past and present energy supply and demand trends but concedes that there is any number of alternative scenarios ahead. The halving of the oil price in 2014 caught the great international oil institutions by surprise, not only the IEA but OPEC itself.

What Lies Beneath

For four or five centuries, Western intellectuals have assumed the superiority of Western culture and values, and of Western economies and industrial dynamism. The reality was there for all to see – a growing technological advantage swelling into a vast industrial revolution fuelled first by coal and steam and then, for the last century or more, by oil.

The oil would come first from a variety of sources round the world, from Baku and Pennsylvania, from Romania and Siberia, but then increasingly and overwhelmingly from the Middle East. That was where the massive oil reserves lay; that was the key region and the key source. Coal, imported and indigenous, would continue to fire great industries and power stations; gas would play its part; uranium would be mined and sent to burn in an expanding nuclear sector. But oil was the dominant fuel and the Middle East was the dominant source – and an area of central concern to the oil-dependent Western powers.

A sequence of tumultuous developments have now shattered this familiar pattern, overtaken the mass of assumptions gathered behind it and built upon it, and indeed shaken it to destruction.

The first upheaval is the rise of the so-called developing worlds of Asia, Africa, and Latin America, driven by the information revolution, the microchip and all its associated and consequential technologies, bringing into the global system three billion new capitalists, new megacities, and huge new economies. Industrialization – of a new kind – ceases to be a Western preserve. The East and South, not the West, have started to set the pace.

That has been the unfolding scene over the last thirty years, although many Western opinion-formers and policymakers have been dismally slow to grasp the change of world balance and its consequences – and still have not done so in some quarters.

Now we come to the crunch, the moment of realization. The specific energy consequence is now plain to see – almost all the growth in world energy consumption, now already and for decades to come, is taking place *outside* the West, outside the OECD countries, and in the transforming new giant economies of the East and South. The great issues of both energy and climate, once synonymous with the West, cease to be a European phenomenon. The action and the decisive changes will be elsewhere, almost regardless of European or Western policy.

Between now and around 2040 – over the next quarter century – the demand for oil in the so-called developing countries ('so-called' because many parts of them are far advanced industrially) will grow by about 28 million barrels a day. It could be less as newly industrializing societies leapfrog stages of modernization and move straight to maximum energy efficiency and consequently slower actual growth in energy demand. Even so, it will be far greater than anything hitherto, as billions of people enter fully into the consumer era. Meanwhile, the demand for oil in the OECD advanced industrial economies will fall by some 7 million barrels a day. So a net world demand increase of 21 million barrels a day is predicted to add to today's 90 million barrels a day – totalling 111 million barrels a day in all.

Production increase plans in 2015, not just in American shale oil but in conventional oil recovery and expansion in many countries, will far exceed that level, indicating that surpluses, and therefore weaker oil prices, may be here to stay for a while, in turn dragging down world gas prices and creating a long period ahead of oversupply.

Oddly, the full and massive implications of this shifting pattern hardly feature in many energy projections and forecasts. If low prices persist, as they may, then the board will be shaken and a whole new pattern of winners and losers will appear. We will look more closely at exactly who they are or may in the pages ahead.

Enter Climate Issues. Enter Shale.

The second great upheaval has been the green commitment, already referred to, swelling up over the last ten to fifteen years into the initial but now wavering determination by governments and vocal policy proponents round the world to address climate concerns. The new priority became to develop non-fossil fuel alternatives, phase out fossil fuels, transform both energy production and consumption patterns, replace whole power systems, and bring on the low-carbon age. This was to be THE great global transition to a new post-oil age.

But the challenge to the whole project, to the renewables, and to the low-carbon crusade is cost – and that cost rises as fossil fuel prices fall. Particularly challenging will be the mountainous costs of new, large, state-of-the-art nuclear power stations – some of the most complex and expensive structures ever built by humankind (see Chapter Thirteen).

The third upheaval – much the most recent – resisted by some, not even acknowledged by others (and again directly challenging parts of the green trend), has been the immensely rapid adoption and development of new or improved technologies for gas (and oil) recovery which have succeeded beyond all predictions and dreams in the US and may yet sweep the entire planet. This is so-called fracking, or the deep hydraulic fracturing of shale rocks, combined with new skills and methods for horizontal drilling and pipe insertion, bringing into commercial range new fossil fuel reserves on an undreamt-of scale and from an unsuspected variety of sources and locations.

It is this third energy 'earthquake', upsetting all past calculations – a phenomenon barely a decade old – which alters the whole world energy scene yet again, sets the globe spinning on yet another energy revolution, sends shocks through world oil markets, has major price implications, increases price volatility, shifts wealth and power even more swiftly into new hands, upsets, to a degree, low-carbon assumptions, and dramatically alters the geopolitics inherited from the twentieth century. And it is this which changes the mood in the Middle East, and makes Arab leaders wonder whether somehow this threat to their predominance can be checked.

The world now looks out on an energy scene which is fast becoming very different from what was predicted, planned, or hoped for. The totality of carbon emissions continues frustratingly to rise, taking humanity with awful inevitability onwards and above the global warming levels – 2°C above current levels by 2050 – where weather and climate turbulence are said by many experts, but not all, to start becoming 'severe, pervasive and irreversible', and really frightening. The 'safe' level to be aimed at for carbon emissions by 2050 is calculated by the UN's International Panel on Climate Change to be 40% below 1990 levels. The actual level on present trends, say the BP economists, will be 25% above 1990. Furthermore, add the pessimists, even if all the pledges made by governments in Paris in 2015 to cut their emissions are fulfilled, carbon levels in the atmosphere will still be rising dangerously, leaving global temperatures at nearer 3.6% above 1990 levels rather than the 'fatal' 2%.

More coal-burning, more gas and oil being sucked or released out of the ground in more places than ever before, unchecked population increase, and the huge and wide paths to growth and development in China and India, covering a third of the world, are all marching towards the wrong destinations – the ones that policy, preaching, and prayers are desperately trying to avoid. What ought not to happen, what must not happen, say world leaders, is happening. Debate swirls, couched in both scientific and quasi-religious terms, about what the threat really is and what is needed to meet it, while below the radar of public and policy debate, something very different is occurring. Governments may rage and regulate, but the carbon expansion rolls on.

CHAPTER FOUR

Low-carbon Paradise Postponed?

The trilemma failures: 1. Global carbon emissions set to continue rising – how gas cuts carbon – American success but world failure – targets certain to be missed – the dangers of dogma – nuclear power hopes dashed with German opt-out – coal prevails in Asia but also in Europe – the limitations of carbon capture and storage – Europe's 'example' strategy dead.

'Spurred chiefly by China... and India, the world spewed far more carbon pollution into the air last year than ever before, scientists announced Sunday as world leaders gather to discuss how to reduce heat-trapping gases. The world pumped an estimated 39.8 billion tons (36.1 billion metric tons) of carbon dioxide into the air last year by burning coal, oil and gas. That is 778 million tons

(706 metric tons) or 2.3% more than the previous year.'[1]

'Europe should only push ahead with its planned cuts to carbon emissions if the rest of the world agrees to a global climate change deal at a crunch summit in Paris next year, according to the EU's energy chief. 'If there is no binding commitment from countries as India, Russia, Brazil, the US, China, Japan, and South Korea, whose governments are responsible for some 70% of global emissions, I think it is not really smart to have a 40% target', the EU's outgoing energy commissioner, Günther Oettinger, told an oil and gas conference in Brussels.'[2]

Let me begin this chapter with a little piece of good news for the green cause – in 2014 the USA, not known in the past for conscientious climate policy, not only achieved the world's largest increase in oil and gas production, but also recorded the largest decline in CO_2 emissions worldwide. It was not particularly welcomed or even acknowledged by anti-fossil fuel campaigners. This was because its cause lay in a massive displacement of coal in power generation by cheaper shale gas – one fossil fuel replaced by another, although emitting about half as much CO_2 when burned in a power station.

Back across the Atlantic in the EU, by contrast, carbon emissions fell only marginally and mainly as a result of the economic recession. Actual carbon intensity – the amount of carbon emitted per unit of energy produced – rose as Europe's economies replaced gas in power generation with cheaper coal, imported from the USA among other sources, notably Russia.

1. Seth Borenstein, *Associated Press,* 21 September 2014.
2. Arthur Neslen, *The Guardian,* 25 September 2014.

Add in the carbon content contained in Europe's swelling volume of imports from countries which did not, and do not, comply with the same high standards as European firms, as well as carbon emitted from other non-power sources, and you get the true picture – Europe's contribution in combatting climate change has been minimal and may actually now be negative.

Worldwide, despite the encouraging American emissions reduction (albeit from very high levels), the picture is far more uncertain. No one quite knows whether worldwide emissions are currently rising, stable, or, as one recent UN report claims, actually falling slightly. But for the future, BP estimates that CO_2 emissions globally will keep on rising over the next twenty years at a rate of at least 1%. This is by far the most likely scenario. Population growth alone makes it so. Thus, there is not the slightest hope of the 40% reduction by 2050 to 'save the planet' – or not at least from changes in the fuel mix.

So why, despite all the passion, all the Paris promises, all the penalties, all the speeches, all the marches, all the unfairness, all the suffering, are global emissions still inching upwards towards the levels which may one day bring terrifyingly increased climate violence, not to all but to many? Global carbon dioxide emissions, or their equivalent, are currently running at around 50 billion tonnes per annum. On present trends they will rise to 68 billion tonnes per annum by 2030. Yet climate scientists insist that if the planet's warming is to be kept at or below 2 °C above pre-industrial levels, if the all-important 2050 end goal is to be achieved, a maximum of another 1,000 billion tonnes (1,000 gigatonnes) is the very most that the already carbon-laden atmosphere can take without catastrophe.

Work it out and assume optimistically that emissions can be kept to the 50 billion annual figure over the next thirty-four or thirty-five years. The numbers cannot possibly be made to add up. Can there be the slightest chance of keeping within the emissions limits?

The answer is both complex and simple at the same time. The projected growth of emissions arising from production and industrial activity is of course almost entirely in Asia, in Africa, in Russia, and in South America. Even if the whole of Europe was cutting carbon emissions – which, as we have seen, it is not when carbon from both consumption of imported goods and domestic production is taken into account – the effect would be statistically negligible. The impact could only be through example – encouragement to the big emitters to take much tougher action. But the example is not being followed. European decarbonization policies, even if they were truly working, would add little.

But wherever the source, the stark and basic conclusion is the same. The policies being pursued are just not working on a global scale. The situation is akin to a child at the steering wheel of a car. Like policy, the vehicle zigzags from side to side, smashes into obstacles, accelerates much too fast, seriously endangering passengers and bystanders alike, even hurting them badly. The journey was the right one to set out on but it is ending in a car crash.

The green cause seems so right and righteous and so widely shared as an ideal and sustainable way for the planet to go that there has been huge reluctance to admit that the policy structures underpinning the cause might be at fault. Criticism of the policies has all along been dismissed as denial of the problem or an attack on the green vision. False polarity has dumbed down the debate and the supporting analysis.

Yet the evidence of failure is now so clear and so worrying for those who accept the strong possibility of climate violence but see the trend worsening all the time that the moment has come to be open and frank about WHAT IS GOING WRONG. We therefore need to examine closely the reasons for such an obvious lack of success, analyse why decarbonization strategies are just not working and why green policies are leading to anything but green outcomes. It will be interesting to see whether it is possible to conduct this sort of questioning and examination without this author immediately being denounced as a climate denier, as an irredeemable sceptic, and as someone who, in the chillingly unscientific words of Sir Paul Nurse, President of the Royal Society and an eminent scientist himself, should be 'crushed'. Probably not.

In the matter of energy and climate change, we are dealing not with facts but with convictions and beliefs. The 'tail' of dogma has long since been wagging the 'dog' of common sense. What is perfectly and visibly obvious is simply being ignored – that world CO_2 emissions are rising, that on present trends 2050 targets cannot possibly be met, that 'dirtier' coal-burning is spreading round the world, not shrinking, that the subsidies required to support greener energy sources, instead of being contained, are becoming insupportable, that adaptation to growing climate extremes is not being funded or implemented at anything like the required pace or level, that the European net contribution is minimal, the only hope being that it will lead by example.

The high priests and oracles of climate concern have simply not yet been able to broaden their thinking to include other confessions. For them, any suggestion that it is possible to accept the case for combatting global warming without obeisance to the full current battery of 'green' policies, programmes, targets, carbon budgets, and ambitions is tantamount to heresy. No land, no space lies in between. Thus, their faith is in a very early and absolutist or dogmatic phase. It requires total, unquestioning commitment. To challenge its effectiveness in leading the way or achieving results is sin and is deserving of damnation – being lumped together with the lost souls, the ultimate 'deniers'.

The story is rooted in a misunderstanding which is indeed childlike. Determined to check world carbon emissions growth, and somehow check global warming to less than 2°C by 2050, the campaigning climate strategists reasoned that *all* fossil fuels – coal, oil, and gas – and the entities extracting and purveying them were the main enemy. Therefore, if their use could be phased out worldwide, via both high prices and high taxes, and if low-carbon 'winners' could be encouraged, subsidized, and incentivized to the right degree, the battle against rising emissions globally would begin to be won. Europe's example would lead the other main carbon-emitting countries to sign up to a legally binding deal, bringing colossal net benefits to the European economies. As we shall see in the next chapter, the UK eagerly aspired to these aims and indeed went further, setting itself even more rigorous unilateral targets through the 2008 Climate Change Act – the so-called gold-plating tendency often practiced by zealous Whitehall officials, and, in this instance, supported by all the political parties in an awesome consensus.

Staying Cheap and Plentiful

But this vision of a world turning away from scarce and expensive fossil fuels has not materialized. Contrary to what most people believe and many experts have forecast, both gas and oil production capacities are now growing at levels and at a pace certainly rivalling the heyday experiences of the twentieth century and which may be unprecedented.

Surplus energy is hitting the world, and oil storage tanks are full to the brim. But the inevitable policy adjustment to these new conditions has not yet taken place. This is partly because the old, discredited theories of peak oil – 'the world is running out of scarce resources...' – linger on in policymaking circles and interested lobby groups. But it is also because unlimited pools of oil and new recoverable sources of natural gas spell lower fossil fuel costs, and these have duly begun to emerge. This, in turn, is bad news for all investment in green and renewable energy sources *at their present cost levels.*

The proviso is crucial. It is saying that if technology, research, and innovation could dramatically lower the costs of renewable and green energy sources, the balance would change sharply. The extra cost of 'green' energy would be narrowed and overturned to the point where green power would begin to mean cheap power. It would imply that by far the best path to a decarbonized world would be through maximum direction of efforts and resources to cheapening new technologies in the production of electricity from wind, solar power, the sea and tidal power, river flows, and a dozen other energy sources. Notable amongst these would be a downward breakthrough in the costs of storing electricity, for this would overcome at a stroke the huge disadvantage of intermittency of power supply, especially from wind power.

This would be the opposite of most current policies towards renewables – certainly in Europe. Current policy has been to subsidize *existing* renewables operations and projects, built with *existing* methods and using *existing* technologies. At the best of times, this spells big subsidies, paid for by high taxation to finance government support, plus high prices for consumers, over and above what it might have cost to produce electricity efficiently from burning coal, oil, or gas.

But with coal getting ever cheaper, as massive new deposits are opened up, ready either for deep mining, surface mining, or gasification underground, with oil prices sagging downwards and with gas in some markets at record low prices, with the effect spreading to other markets, the whole subsidy structure soars in size and cost. (See the next section on the second of the three 'trilemma' failures and on unreached goals – affordability.)

With governments everywhere feeling weaker and confronted by shrinking and infinitely more difficult tax bases, subsidies are becoming too large to bear and price-inflating charges on consumers are rising too high for governments to impose or to do so without threats to their own survival. The reactions of both domestic and industrial consumers have truly set in. Exceptions, exemptions, special grants, and compensation create a mass of loopholes in the overall high price structure for hard-hit industries and families, bringing the real politics of decarbonization, in contrast to the rhetoric, almost to a halt.

There had been a hope that a revival of lower-carbon nuclear power would, in due course, fill the gap. Admirers and students of James Lovelock, the interpreter of Gaia, have noted his thinking that low-carbon nuclear power may be on the right path to the intelligent preservation of planet earth and its atmosphere.[3] But here, too, the cost trend and, therefore, the subsidy requirement have gone the wrong way. Chapter Twelve examines all the low-carbon renewable prospects, including

3. James Lovelock, *A Rough Ride to the Future* (Allen Lane, 2014).

nuclear power prospects worldwide, but the dream of cheap or at least slowly cheapening atomic power has not arrived. On the contrary, the overall costs, capital, and running of nuclear power have risen remorselessly. Technological breakthroughs which would significantly cut costs and simplify construction have not yet occurred. When they come they may bring us much smaller and safer nuclear plants, built much more quickly on a modular basis – a prospect examined in Chapter Thirteen.

For some green sources, things have gone better. Solar power costs have fallen dramatically. In the sunnier climes of the Middle East and North Africa, ultracheap photovoltaic cells and panels imported from China were until recently beginning to make solar power competitive with oil-fired electricity generation – that is, until the oil price began to slide, opening up the cost gap again. But cheaper solar power could be within reach and would certainly gain a foothold if methods of electricity storage could themselves be reduced in cost and size, while hydropower is also a source of cheap electricity round the world, even when all environmental costs are taken into account.

On some American onshore sites, electricity from wind power, taking into account all related system costs and even allowing for intermittency, have come near to competing with coal and gas sources. American offshore wind costs, by contrast, remain at three or four times the cost of conventional electricity.

But on the nuclear front, cost reduction has yet to begin. On the contrary, costs keep rising and rising. In part, the upward cost trend has been driven by rising safety requirement features and demands for strengthened anti-terrrorist features.

But the biggest cost is delay. The Daichi Fukushima disaster has greatly fuelled political fears, spelling not only ever-longer delays on projects in the face of popular resistance but outright withdrawal of government support. The German case, examined in Chapter Eight, provides the classic and devastating example of this process.

With nuclear power being closed down in Germany by edict of Chancellor Angela Merkel, the contribution that it could have made to reduced carbon emissions has vanished. Despite frantic efforts to encourage other green energy sources, carbon emissions in Germany are rising. The policy is producing its obverse – and with heavy collateral damage in terms of sky-high energy prices, rising fuel poverty, and a severe loss of competitiveness in German industry.

Coal wasn't meant to fill the gap

If one looks for the most visible demonstration of policies gone awry and unintended consequences from well-intentioned measures, the clearest manifestation is the renewed ascendancy of coal – coal of all sorts, in all forms, and from many different sources. The Fukushima nuclear setback was of course unforeseeable, or at least unforeseen except by a very few. But it has inevitably driven up the cost of new nuclear power worldwide and driven down the chances that a nuclear renaissance could contribute to the achievement of carbon targets or the security of power supply.

Instead, the gap is being filled and is being consciously planned by investors to be filled in the future by more fossil fuel-based electricity generation, by gas as it grows increasingly plentiful, but even more by the one fossil fuel which is supposed to be on the way out, the one which has by far the largest carbon content and which contains numerous other 'dirty' particles – namely, coal.

There are now reported to be no less than 2,177 coal plants being planned worldwide. 557 of these are under construction.[4] Of course, not all the rest will get finished, or even started, but even if a quarter of them get built, the impact on global carbon emissions will be huge. Carbon budget limits deemed necessary to keep the global warming threshold increase of 2°C by 2050 will be shot to pieces.

Most of these new coal stations will be in China, India, Vietnam, and Indonesia. In Indonesia they are talking about a 'global coal renaissance'. But five of the richer nations – Britain, France, Germany, Italy, Japan (all G7 members) – also plan to burn more coal over the next five years.[5] The exception to the upward trend is the United States, where the switch from coal to gas has been massive – and driven initially not by government policy but by shale technology and innovation, leading to plentiful cheap gas. But US coal-burning starts from a very high level. Until recently, 47% of American electricity continued to be produced from coal-fired stations. Now some of 'king' coal's subsidies and protection have been removed by pressure from President Obama, and a swathe of penalties and regulations on all US coal activity have been brought in.

The immediate effects have been to drive America's big coal companies to the edge of bankruptcy. Alpha Natural resources, a former giant company, has been forced to declare itself bankrupt, while Peabody Energy, once the greatest coal

4. According to Platts WEPP and CoalSwarm analysis.

5. This is despite the statement from the UK Energy Secretary, Amber Rudd, that all unabated coal-fired power stations in Britain will be closed by 2020!

company of all, has seen its shares collapsing from $70 to a shade above $1. Arch Coal is facing the same prospect.

So at least in America, still the world's biggest economy, if not in Asia, coal is gradually on the way down, if never entirely out. But even then, is the fraught situation facing the coal industry quite what it seems? Why, for example, is George Soros, the proverbially shrewd and far sighted investor, actually buying large blocks of shares in these companies? Soros is the master of market oscillations, always acutely aware that financial markets, in particular, invariably overreact. He clearly sees a way back for coal and he is probably right.

Why is coal – the big emitter of carbon, as well as a more immediate and serious polluter – not being penalized out of the energy picture, as almost all low-carbon campaigners insist should happen and see as a necessity? The answer is that, except in shale gas-rich America, it is still very much a cheaper fuel to burn than gas, even with carbon taxes and other levies and disincentives added.

In Europe, where gas prices have so far remained about three times as high as in America, and where cheap coal imports are on offer from Australia, from American coal producers seeking new markets, from Central Europe (the uniquely carbon-laden and polluting lignite coal from Czech and Slovakian suppliers), from Russia, and numerous other sources, this fuel is inevitably the attraction. Existing and quite new gas turbines are being closed down for being straightforwardly loss-making. Their owners cannot carry on bleeding. They cannot compete with cheap and plentiful coal or with subsidized wind supplied at marginal cost – when the wind blows. Coal may have been forced out of the US for the time being, but not elsewhere.

On an even vaster scale than anything in Europe, Australian coal-carriers are moving across the South China Sea, feeding dozens of new coal-burning stations opening up across China – a new one almost every week. India's ports are jammed with inward coal carriers – again from Australia but also from Papua New Guinea and Malaysia. In one year alone, 2014, China *added* coal-burning capacity to generate its electricity needs on a scale which was double the size of the UK's entire 'fleet' of coal stations (which produce annually about 20 gigawatts of electric power).

In Japan the advance of coal is even more dramatic. Work is under way on no less than forty-three huge, new coal-fired power stations with a capacity of 21,000 megawatts (about two-fifths of the UK's entire electricity output). Leading the new march of coal is an immense project in Akita Prefecture, being built by Kansai Electric and the Marubeni Corporation. At 1.2 gigawatts, the size of a large nuclear station, it is one of the biggest coal-burning stations ever. India, as mentioned earlier, is committed to doubling coal production to feed its growing string of coal-fired power plants.

Pakistan is following suit. With massive Chinese support amounting to $37 billion of funding, new coal-fired stations are being built across Pakistan, ten of them due to generate 7 gigawatts of electricity in the Thar district. The whole Thar Coal and Energy Project has particular significance because this is precisely the scheme which the World Bank refused to back on the grounds that it involved environmentally unsustainable characteristics. So here it is being backed instead by Chinese finance and no doubt in due course by the new Asian Industrial Development Bank which is being set up, amidst strong American protest, to finance precisely these sorts of energy projects in the mid-Asian region.

South Africa is aiming for a major coal expansion – both in production and power plant supply. Indonesia, Vietnam, and

South Korea are on the same path – all planning to increase their coal-fired capacity. Bloomberg Intelligence analysts come up with the staggering figure that, taken together with Japan, these countries plan *additional* generation from coal of 204 gigawatts by 2019 – four times the UK's total power output.[6]

The Indian Minister of State for Power, Coal, New and Renewable Energy, Piyush Goyal, put it bluntly: 'India needs to dig twice as much coal as it does today if it is to meet its soaring energy demand'. By 2019 it is expected to be consuming 2 trillion units of electricity annually with 1 unit equalling 1 kilowatt hour. Describing coal as 'an essential input for power', Goyal said: 'I see Coal India production doubling in the next five years. It makes about 500 million tonnes hopefully this year. We [will] do a billion tonnes in 2019'.[7] The Indian scene is becoming still more challenging. 'The fundamental objective of our country is that coal availability goes up, and there is 24/7 power for everyone', says Sutirtha Bhattacharya, chairman of state-backed Coal India. By 2025 India will overtake the USA to become the world's second largest coal consumer – and, shortly after that, will overtake China and move to first place.

Meanwhile in Germany, the share of electricity generated from coal-burning is back at levels not seen for quarter of a century past. In 2013 the Germans opened more new coal-fired plants than at any time in the past twenty years. As German nuclear power began to wind down to total closure by 2020, the Chancellor Angela Merkel pronounced more coal-based generating stations as 'indispensable'. Her federal economics minister Sigmar Gabriel goes further. Announcing that Germany was planning to withdraw from its previous 2020

6. *Bloomberg,* 17 April 2015.
7. Climate News Network, 14 November 2014.

goals for cutting CO_2 emissions, he explained, 'We cannot exit from coal power overnight.'

Germany imported around 50 million tonnes of hard coal in 2014 which was just below the previous year's record of 51 million tonnes, the coal importers' lobby, VDKI, said in a statement. Coal generation obstinately remains the backbone of German power supply in a country set on moving away from nuclear power and favouring renewable energy over fossil fuels. Between January and September 2014, the country used hard coal for 43% of coal generation, of which 17% was black coal and 26% the other really dirty kind, brown lignite coal, so industry statistics showed. With power prices historically low, generators were simply and inevitably opting to burn more coal instead of more expensive gas.[8]

From the point of view of the energy investor, the choice is loaded and obvious. As George Soros clearly believes, even in the US the coal business will somehow remain profitable. Round the world, coal-fired generating stations remain a good investment for the foreseeable future (and are, as a result, being constructed all over Germany, although not in the UK– see later chapters). Big, new efficiencies in coal-fired generation enlarge profitability further and promise lower-carbon emissions per unit of power output.

By contrast, gas turbines are under assault from three sides. Coal is taking their markets; carbon and other taxes are raising their costs; and the subsidies are going to their direct competitors – wind farms and solar installations. Money into gas-fired capacity currently makes no sense for investors. If existing gas turbines are making no money selling electricity, who in their right mind will invest in building new ones?

8. *Business Recorder*, 5 December 2014.

Aware of this dilemma, the British authorities, amongst others, have brought forward elaborate belt and braces devices to coax investment into new-build gas plants. Whilst calling for all coal-burning power stations to be closed – or at least the ones without any carbon-diverting devices – the urgent priority has become somehow to induce investors into new gas-fired plants. So-called capacity payments are being offered to anyone who will build and have ready-to-operate gas turbines, even if no gas is sold. They will be paid to exist – and at an attractive rate. And, at the same time, the British have introduced a uniquely high carbon tax on coal-fired generation – between £16 and £18 per tonne of carbon emitted – as against about £5 on average across the rest of the European Union.[9]

For the energy investor this ought to sound a dream come true – revenue guaranteed in exchange just for existing and maintaining a gas-fired plant in standby readiness. Under a kind of reverse auction process, the companies which will invest and build new capacity and keep it always available for the lowest promised income stream of payments from government will get the contracts and the deals.

Why is there not a rush for the offer? Because even bigger 'deals' with bigger subsidies are available for owners of wind farms and other renewable constructions, and, ironically, for coal plants as well. The rules are permanently tilted against gas. Only the very brave and the foolish are going to fall for this one – even if their shareholders let them. (See Chapter Nine for much more detail on the extraordinary tangle into which British energy policy has led itself.)

---

9. This is the total sum UK coal burners now have to pay and includes the EU-wide charge under the Emissions Trading System. The rest of the EU has no topping up carbon tax, and so just pays the minuscule emissions permit price.

The pattern is one of too little, too late. Belatedly, the total opposition of climate policymakers to any extension of fossil fuel-burning has been modified to concede that gas may be 'the transition fuel' on the low-carbon path.

A 2013 study by the Grantham Research Institute for Climate Change and the Environment at the LSE, no friend of climate scepticism – in fact the extreme opposite – reports that shale gas helps to cut carbon 'pollution'. Therefore, the UK should go all out to replace coal for gas power supplies as it has been shown that burning coal causes more smog, soot, acid rain, global warming, and toxic air emissions than extracting shale gas.

Bob Ward, one of the study's authors, said to the *Financial Times*:

> 'There is a very clear logic to using gas as a transition fuel because over 20% of carbon dioxide emissions in 2011, the last year we have data for, came from burning coal.'[10]

The Institute is, of course, scientifically wrong to equate carbon emissions with air pollution. Carbon in the atmosphere is certainly a greenhouse gas but it is not a pollutant or the cause of the heavy and dangerous fogs and smogs that plague Chinese cities and used to paralyse London half a century ago. That job is done by sulphur emissions and other noxious fumes that every government everywhere is sensible to try to prohibit and shut down.

But the case for gas on lower-carbon grounds is strong and well-proven. It is the obvious way to go, as American experience proves beyond doubt. But for Europe it is too late. The anti-gas slant of energy policy and political strategy cannot be reversed

10. *Financial Times*, 19 March 2013.

just like that. The battle has been lost, and lost mainly and ironically to high-carbon coal.

In sum, this is the energy story that is not being told – or, in the case of the most earnest low-carbon proselytizers, is being ignored. Coal may gradually account for a diminishing percentage of total world electric power. But that total will be vastly greater by 2040 or 2050. And the projections for increased world coal-burning will unavoidably be vastly greater within it. Having risen by 200% since 1980, coal use will rise another 100% from now until the late 2030s, although contributing by then about 30% of a very much larger total world electricity output.

This is the ugly, unwanted, but inevitable reality which has to be set along side the 'inevitability' of a low-carbon world which the green empire proclaims. And nothing done in the West or in Europe or in the UK or decided in Paris is yet in sight to prevent it.

However brave the example in Europe, however strict the unilateral targets in the UK, however heavy the penalties on fossil fuels in the West, coal-burning will increase and with it CO_2 emissions.

Is there any way out? Yes. Technological advance on a Marshall Plan- or Manhattan Project-scale could in theory be applied at the source of the problem – coal-burning in Asia. It could ensure that more coal is burned so long as and if the resulting emissions could possibly be curbed.

But it is a big and very expensive 'if'. In the UK, over a billion pounds of taxpayers' money was earmarked in the attempt to get carbon capture and storage working commercially.[11] Round the world, governments together have spent about $24 billion. The

11. Although this whole programme has very recently been closed.

results so far have been paltry, although there may now be much cheaper ways of utilizing carbon rather than piping it away to storage (called carbon capture and utilization – CCU).

There is just a chance that if much more than this was spent and put behind a huge technology push, there could be some restraint on the galloping horse of coal expansion. It might not be too late to do this, but the funds would of course have to come from subsidy funds going to other renewable energy sources. This would require a fundamental switch in energy and climate policy priorities and in the minds of policymakers. It would also require great political bravery. Of that, at least in Europe, there is not the slightest sign.

The Consumer Misbehaves

For those hoping that, nevertheless, coal stations will be closed (by EU directive) and lower-carbon emissions overall will follow, there is another disappointment and another realization ahead – namely that policies aimed in one direction produce outcomes in quite another. Thus, while energy intensity in production in most of Europe has undoubtedly fallen, and indeed during the austerity years the production of emissions has been notably trimmed back (about 4% below 1990 levels), the consumption of emissions per head has greatly increased (in the UK alone by around 28% per head over the same period).

Why? Because imports of manufactured goods into Europe have soared. Manufacturing production, driven away from Europe and especially Britain, by high costs – including high energy costs – has taken place in China or India or Malaysia where restraints on carbon emissions are weak. This is the so-called carbon leakage phenomenon in action.

What kind of example to the rest of the world is this? The first answer is not much, and the second answer is that it makes very little difference to the major world emitters either way. All along, the countries with the major growth of emissions – notably India, China, and Russia – have been following their own agendas – not necessarily agendas of denial but nevertheless ones supremely uninterested in Western or European examples of any kind. In as far as any interest at all has been shown in Western policy, it has been to assert that it is the old industrialized West that has put most of today's global gas emissions into the atmosphere and the West should therefore pay up in full for the extra costs in the developing world of moving from cheaper fossil fuel power generation to expensive low-carbon alternatives.

Since the sums are calculated to be astronomical (China's opening bid was $200 billion, but a figure of $400 billion is now being mentioned), it was until recently assumed that this would simply not happen. How could developed country governments, all feeling the pinch on their spending in an age of austerity, possibly open up a cornucopia of this size, with the possibility of it being open-ended and leading to even bigger transfers of both funds and technology?

Yet recently the concept has been given a small puff of new life. President Obama hinted that more help for poorer countries facing extreme weather conditions might be on the agenda when the outcomes of the major Paris conference on climate change in November 2015 are followed up. Some kind of modified 'deal' could thus be emerging whereby the industrialized and advanced nations pay up in exchange for firm commitments from the new giant emitters (i.e. China and India) that they will curb emissions, slow down coal-burning etc. In the money-laden words of the Indian Prime Minister, Narendra Modi, 'the focus must shift from climate change to climate justice'.[12]

Examples and Demands

Hitherto, the argument and hope of climate and energy policymakers in Europe has been that the major emitting nations will somehow 'come round', modify their demands for Western compensation, and agree to legally binding limits on their carbon emissions or at least firm undertakings that limits will be met in due course. Mr. Modi's words are therefore ominous for Western governments on tight budgets and trying to reduce debts. The reality is that neither China, India, nor Russia have any intention of following the Western 'example'. Each country may recognize responsibly that climate warming is a serious danger. But these three are following their own paths. They are not swayed by any example. Any swaying may have to be done by colossal sums of money – and may not guarantee results in terms of reduced carbon emissions even then.

Thus, the new Indian Government has made it crystal clear that coal-powered electricity will continue to expand because coal is the cheapest power source and coal makes development and economic growth possible. In a country of still vast poverty, its reduction, underpinned by the essential ingredient of cheap power, must come first. The Indian Prime Minister, Mr. Narenda Modi, has reiterated this position with firmness and clarity.

China, too, has its own priorities. Its leaders take a responsible and concerned position about carbon emissions but their

12. *Press Trust of India*, 3 September 2015.

pathway is not the West's. Both at the December 2014 climate 'summit' in Lima and in bilateral exchanges with President Obama, China undertook to secure a peak in CO_2 emission (of 12 gigatonnes of CO_2) by 2030.

But that is not by following anybody else's example. On the contrary, it is specifically avoiding the example of Western economic growth and aiming to follow a carefully moderated growth pattern embodied at the Fourth Plenum meeting in November 2014 in the phrase or slogan, 'the new normal', and in the single word, 'enough'. The Chinese are saying, in effect, that the debt-fuelled, hectic, consumer-driven growth patterns of the West are not for them.

Nor will they bind themselves with promises and targets which cannot be achieved and which in practice distort energy investment in highly negative ways. Instead, they calculate that an emissions peak in fifteen years' time can be reached not by rejecting fossil fuels but by burning them more efficiently (including from an enormous fleet of new, 'cleaner' coal-fired stations). They also forecast that much greater efficiency in energy use can be achieved. That will be the Chinese way.

Innovation to the Rescue

The drivers along this alternative path will be not penalties but innovation – innovation in primary resource production (fossil and renewable), innovation in generation and transmission, innovation in every conceivable form of use both industrial and domestic.

Heavy concentration on new technologies to promote energy use efficiency and new methods of electricity generation, whether by coal, gas, oil, or nuclear power, have been given priority. Dozens of new coal-fired stations will be built but they will be vastly more efficient and thus reduce emissions per unit of output. Depollution of all fuel burning will have high importance to meet the immediate smog threat to Chinese cities. Efforts to cut the cost of low-carbon alternatives, notably solar power, will also be unremitting. But this will be China's way at China's pace (see the full remarkable story at Chapter Thirteen).

The confusion in the minds of, for example, BBC commentators who say that carbon dioxide is the same as and, indeed, somehow the cause of air pollution in Chinese cities, is comical. (This was referred to in the Introduction.) Of course, CO_2 has many properties, but causing air pollution is not one of them. That comes from quite different and increasingly preventable sources as the Chinese Government recognizes – including coal-burning. It is preventable, as firm British action proved in the 1960s with clean air legislation which made the dreaded London smog a thing of history. Most of the appalling smog in Chinese cities in fact comes from transport and from coal burned in domestic hearths.

Russia will continue to burn both coal and its plentiful oil for power without the slightest concern for the Western example, arguing, ingeniously, that it is doing its bit against climate warming by soaking up carbon in its immense forest areas. Trees are Russia's green policy.

In the Middle East, the example theory makes even less impact. Per capita oil consumption is soaring as highly subsidized oil is used increasingly to generate power for the region's big needs for power for air conditioning. One estimate is that Saudi Arabia will consume so much oil that by 2030 little or none will be left for export.

In Japan the abandonment of official 2020 decarbonization targets was announced in the autumn of 2014. The previous aim had been to reduce emissions between 1990 and 2020 by 25%. Now the realistic assessment became that they would RISE by 3% over that period. Between 2005 and 2020, the hope was that they would fall slightly – by 3.8% – but that of course was a hope. In practice the have continued to rise – up 1.3% in 2013 and more in 2014.

The Japanese authorities are clearly trying to nudge public opinion into facing up to the appalling cost of keeping their nuclear power sector closed. They could now be slowly succeeding with the first few out of forty-three stations being allowed to open up. Closure has involved importing oil and LNG on a colossal scale at a cost of ¥9 trillion a year (about £30 billion). They also believe that money is much better spent on research into lower-cost energy sources and technologies of every kind rather than subsidizing high-cost renewables. At the same time, the Japanese have authorized several new coal-fired stations which, they claim, will greatly increase efficiency in coal-burning and deliver much more electricity per unit of coal (up to 45%), thereby reducing emissions substantially.

This saga should be a warning to all those who try to rush decarbonization impatiently by crude and politically insensitive means. They should draw the same warning from what is happening in Germany, where an ill-managed attempted dash to renewables targets and emissions reduction has resulted in sky-high energy prices and a big leap in coal-burning for electricity, with emissions going up instead of down.

Even the German Government, for a long time the most dedicated authority in Europe to low-carbon goals and architect of the great 'Energiewende' push to green power (of which more in Chapter Eight), has now withdrawn its 2020 climate change targets. The self-imposed requirement to cut emissions by 40% compared with 1990 levels has been abandoned.

'It is clear that the 2020 target is no longer viable', said the federal economics minister Sigmar Gabriel. Low emissions gas turbines are closing down while American, Australian, Russian, and Slovak coal and brown lignite pour in.

His words were further reinforced when the German Government wrote pleading letters to the Swedish mining company, Vattenfall, begging it not – repeat not – to abandon plans for opening new coal mines in Brandenburg. We shall see in a later chapter how Germany's energy policy flaws drove it to these extraordinary contortions, heading in exactly the opposite direction to its goals of hyper-rapid decarbonization and the greening of the whole German power sector.

Why are they not following us?

In short, the 'example' strategy is inoperative – dead. Other forces may be at work giving the appearance (a false one) of reducing carbon emissions in Europe (not least lower economic growth). 'Deals' and commitments have been made. But the overall contribution to global emissions reduction is minuscule and not seen as an example at all. Some realists have quietly pointed out that Europe itself may anyway be set in the longer term on a carbon-loaded future, whatever the short-term downward trend, as new towns, roads, airports, and rail systems renew Europe's often rickety infrastructure.

A final twist to the 'example' story is that a major European device for cutting carbon emissions, tried in several countries outside Europe as well, has proved utterly ineffective. This is the 'model' of making permits for the emission of carbon tradeable through the European Trading System (ETS). The price of these permits was intended to be high enough to deter the big emitters in industry from clinging to old methods and incentivize them to adopt new low-carbon technologies in their factories and plants. As an economist's theory, it looked terrific – on paper.

In practice, with economic growth slowing to a halt in most of Europe, carbon permits have not been needed. Companies being issued with them have become sellers. Companies and firms which have high emissions have only had limited demand for them. The fact of there being more sellers and fewer buyers has meant price collapse. The plan was for a price of about $30 a tonne of carbon. It is now about $5 – no disincentive to higher emissions at all.

For a while, this particular conclusion – that this emperor had no clothes – was regarded as an appalling heresy which could only come from the mouths of climate deniers. But in early 2013 a large group of 130 impeccably 'green' and concerned scientists and campaigners came together to demand the abolition of the European Union Emissions Trading Scheme. Their reason? Simply that the ETS had made no contribution at all to reducing greenhouse gas emissions. It had generated, they pointed out, large windfall profits for certain utilities and had actually led in some cases to increased emissions, thanks to offset projects and the equally valueless Clean Development Mechanism.[13] It was time, pleaded these environmentalists, 'to make room for climate measures that work'!

This has been the final nail in the coffin of Europe's inefficient decarbonization strategy. This part of the approach to the trilemma has failed.

---

13. The CDM allows emission-reduction projects in developing countries to earn certified emission reduction (CER) credits, each equivalent to one tonne of CO_2. These CERs can be traded and sold and used by industrialized countries to meet a part of their emission reduction targets under the Kyoto Protocol.

CHAPTER FIVE

Goodbye Cheap Energy

The trilemma failures: 2. Europe goes for expensive energy – the true costs assessed – the backlash of public protest – no mercy for the poor – the billions who lack cheap power – an odd political stance on the Left – the real low-carbon path.

Cheap power has become a distant memory in most of Europe and especially in Germany and Britain. A world in which government ministers could hold out the promise of 'cheap power for all' has vanished into history.[1] Only for a brief time in the 1980s was there a period of a few years in which abundant oil and plentiful gas held consumer prices down, and only in France today, with 70% of its electricity still flowing from nuclear power stations, built with amazing rapidity in the 1970s, have low energy bills survived.

1. As promised by ministers at the 1956 opening of Calder Hall nuclear power station by Queen Elizabeth II.

For the rest of Europe, the pattern is one of prices deliberately and persistently increased by governments in the name of conservation and to satisfy EU imposed targets. Affordability goals have been brushed aside. Indeed, energy costs are scarcely mentioned in the EU's policy documents in recent years. Yet the energy price issue is at the heart of the EU's economic future and recovery. For example, for Poland, and not just for Poland, the huge extra costs being imposed on industry by the subsidies for expensive renewable energy are no longer bearable. The threat is to veto recent EU proposals for the target of 40% reduction in emissions by 2030. It would, say the Polish leaders, 'destroy half of Europe's industry'.

For the domestic consumer, there has been real pain. In Britain gas and electricity prices have risen, after inflation, by more than 50% in the last ten years. Meeting the European Union targets for reducing carbon emissions – 80% by 2050 – would involve equally large future increases to raise the funds to pay for the gigantic subsidies to renewable energy – notably wind – now required. (See Chapter Nine below for the full details of this whole messy episode.)

The EU carbon targets would mean covering an impossibly large area of Europe with turbines at a cost of €3.2 trillion – most of it to be squeezed from consumers. Over 160,000 new wind pylons would need to be erected by 2020. And all this while the actual costs of *primary* energy supplies – gas, oil, and coal – are actually falling!

Even while this book is being written, the public backlash against expensive power is now beginning, and it is taking many forms.

A vivid example is the rising volume of protest from industrialists throughout Europe, but especially in Germany, against ferociously high energy costs which, German leaders assert, are driving more and more investment to cheaper energy climes, notably the USA, where gas is currently priced one-third lower than in Europe.

Closer to everyday life is the painful increase in household energy bills, driven not by basic energy prices, which have now been falling, but by green levies and taxes to finance generous subsidies to wind farms. This was described by one UK Coalition minister, Owen Paterson, in a carefully worked presentation, as the biggest redistribution from poor to rich since the activities of the Sheriff of Nottingham (Robin Hood's antagonist). The minister was sacked.

To a much greater extent than the standard cynicism of political commentary usually concedes, politicians in all parties are motivated – and have often come into politics – to wipe out poverty. Yet here is a policy specifically designed to hurt the poorest and benefit large land owners and international energy consortia. One the many fascinating twists of modern politics, as it drifts away from the old stereotypes of Left and Right, is that this open attack on the poor received virtually no comment from left-wing parties – for instance, the British Labour Party whose central purpose and commitment had always been to help the poor. In this case, helping the poor was in conflict with policies supposed to reduce carbon emissions. In ways which must have had the great founders of the British Labour movement spinning in their graves, lower emissions won – the final irony being that backing wind farms and the better-off against the poorest may not have been helping reduce carbon emissions at all, or, at best, only by microscopic amounts.

Labour may have 'lurched to the left' since its defeat in the British General Election in May 2015. But it is noticeable that amongst many of the bloodcurdling new left-wing policies being espoused, including wholesale renationalization of energy industries, a halt to wind farm subsidies is not included. Indeed, when under the new Conservative Government a modest proposal was introduced to halt – in due course – various types of subsidy support for any new onshore wind farms, this was vigorously opposed by a Labour opposition that had clearly lost its way.

They were not the only ones. Almost comically, the Royal Society for the Protection of Birds also argued against any reduction in the subsidies for future wind farm construction – overlooking, in its zeal, that wind pylons chop up birds in their thousands!

Cornered by these contradictions, politicians of the Left – and not only on the Left – have turned their attention on the power companies with threats to freeze prices by government edict. But in a beautiful example of the theme to which this book returns again and again – that energy issues are all interrelated – the counterproductive effects of this promise proved twofold.

First, the UK energy companies, apprehensive about a possible price freeze under a possible future Labour government, rapidly brought stocks forward and kept prices well up to ensure they could deliver the 'freeze' level when and if it came about. The second effect was to deter further investment in new generating capacity, and increase the need for costly government inducements to energy investors to stay committed.

With oil and gas prices tumbling far below forecasts, the freeze threat actually worked to keep prices up rather than be cut. The other outcome was to increase the levies and charges needed to keep the power structure intact, hurting the poor even more!

For the very poor, no mercy

For those with broader horizons, there has also been the plight of the one-fifth of the world's population without access to any power supplies to consider. Here, too, one might have expected parties of the Left to demonstrate compassion for those so deprived. Not a bit of it. Remorseless pressure on the poorest communities to stay away from fossil fuels, with Hobson's choice of either leaping directly to super-expensive renewable methods – which cannot be afforded – or sticking with primitive (and toxic) wood or indoor dung-burning methods, has worked directly and cruelly against the interests or progress of the very poorest.

What is missing from these well-intentioned green pressures (apart from compassion) is an understanding that it is lower, not higher, energy bills that will drive the green revolution and lead the big switch from more primitive energy forms. Unaffordable energy means suffering. There is no way of squaring the circle.

The trilemma imposes the highest price of all on the 1.2 billion people, mostly in Africa and developing Asia, who remain without any access to electricity and for whom the prospect of the arrival of cheap and reliable power, far from being brought nearer, is fading away as costly renewables become the only option on offer. Much larger numbers – up to 2.8 billion – may have limited power supplies but still rely on traditional biomass, such as wood and dung, for cooking, often with lethal results.

These are the real victims of the trilemma errors. For these large sections of humanity, a better future and a more sustainable kind of development lie above all through economic growth which in turn depends heavily on plentiful supplies of electricity. Cheap and reliable energy and plenty of clean water, the other imperative of progress, are the key factors in their future water supply being deeply dependent on power supplies, just as power supplies and energy production in most forms depend on water. The symbiosis is complete, and cheap power is the essential binding ingredient.

For these people, the world's poorest, access to cheaper power is tantalizingly within reach. Investment, markets, and technology are fully capable of delivering enormous quantities of power, using cheap fossil fuels and new technology. Hydrocarbons and green innovations can and should work together. For deprived island communities, liquid natural gas can be transported by purpose-built carriers and delivered into quickly constructed island grids and gas turbines. Almost all developing states have the facilities to import cheap coal, either by sea or rail, to fire efficient local plants.

Contrary to widespread assertions, it is this power flow, and the growth which it generates, which provide the resources for better defences against climate change, such as rising sea levels and flooding, which many island states fear. Small-scale systems and mini-grids can go some way to overcoming basic power supply deprivation and relieving hundreds of millions from dirty traditional fuels, lack of light at night, lack of warmth or cooling, lack of energy even for small industries and village pumps, and lack of effective flood protection.

New green technology can certainly play its part. Solar-driven lamps have now been vastly improved and can bring high-quality light to many homes. But sustained development requires organized power supply systems on a larger and more reliable scale. If slums are to give way to proper housing, if unemployment is to give way to regular paid work, tens of thousands of connections need to be made, and efficient power utilities, established, going hand in hand with organized and legally underpinned systems of property ownership.

The lower-carbon path must be followed but it leads through, and not round, more efficient coal-burning and gas-burning. The worst route to lower carbon is the one now being taken, leading to massively increased coal-burning in the bigger world economies, and yet, at the same time, to continued and shocking deprivation in the poorest and smallest communities which remain condemned to twilight kerosene, dirty biomass, and firewood. The best path is to the rapid development of the least expensive and affordable power sources, leading to the economic growth which promotes energy-efficient investment, finances adaptation against climate violence, and eventually opens the route to greater and greener energy efficiency.

CHAPTER SIX

Flickering Lights

The trilemma failures: 3. Unreliable energy – self-sufficiency not always safest – green ambitions weaken network security – the struggle to persuade new investment in new electricity-generating capacity – the expense of carbon capture and storage – other technologies might help.

If two aspects of the trilemma – affordable energy and reduced carbon emissions – have failed so badly, greater success has been claimed by current policy apologists in addressing the third leg of the 'trilemma' – namely, that renewable energy increases national energy security and protection against disruption of external energy supplies. Self-sufficiency is held out as the goal which green power will reinforce. Fuel imports, it is said, are vulnerable to disruption, political threats, blackmailing, and so on. By contrast, green energy is local, secure, and reliable.

But there are two traps in this argument, into both of which it happens that Britain's energy policy has fallen headfirst.

First, the benefits of self-sufficiency in energy supply need to be carefully weighed. In the 1970s Britain was almost self-sufficient in coal yet was brought to its knees by the 1974 miners' strike. This was self-sufficiency of the wrong sort. Besides, in its present technological state, green energy cannot deliver the baseload in electricity that a modern industrial society needs. This would have to be done by a major expansion of nuclear power, which, as we shall learn in Chapter Thirteen, is still very, very expensive,and is taking up to a decade to deliver. In the British case, North Sea oil and gas gave the country a decade or so of minimal net imports of both oil and gas (although, in practice, oil and oil-product trade, involving both imports and exports, took place on a substantial scale).

All experience suggests that the safest pattern of energy supply for big industrial economies comes through diversity – diversity between fuels, diversity between imports and home products, diversity between sources of imported oil and gas, diversity between sources of refined petroleum products.

For cash-strapped smaller island states, the same thing applies. Pushing smaller communities into total reliance on wind or local waterfalls is not necessarily the safest course. Big advances in such areas as floating LNG barges (carrying liquefied natural gas frozen and compressed to a temperature of -162°C) mean that quite small coastal and island communities can have access to gas at very short notice.

But there is a second and more immediate trap lying across the path to low-carbon energy which has already produced dangerously direct, although quite unintended, consequences for the British energy system.

In 2007 the British Prime Minister Tony Blair overrode his industry ministers and committed the UK to a costly EU-agreed target of supplying 20% of power generation from renewable sources (mainly wind and solar energy). The full implications of this for the British situation were either not understood or not honestly shared with the British public. As will be explained in Chapter Nine ('The Coalition Legacy: A Very British Fiasco'), the consequences for the future of Britain's energy situation have been extensive and damaging in many ways, but some of the most serious damage has been inflicted on the resilience and reliability of the whole national electricity supply system.

This has been because the state has had to step in with heavy inducements and subsidies to persuade investors to build wind farms, and then similar subsidies to conventional power station-builders to construct the gas-fired stations necessary to back up weather-dependent electricity. The limited success of this encouragement programme (which will be explained much further in Chapter Nine), combined with EU directives to close down coal-fired stations in short order, and combined also with there being a number of ageing nuclear power stations coming to the end of their life, has eroded the safety cushion of the electricity system to wafer-thin proportions, now about 2 gigawatts as against a more usual 15-17 gigawatts in the past.

A coincidence of very cold weather across Northern Europe with a windless day (as occurs quite often), or with a power station breakdown, a terrorist or cyber threat, or any other fault in the supply chain, could produce serious power interruptions.[1] Green ambitions and commitments have therefore produced out of

1. Self-sufficiency provided no protection against a January 2016 cyber attack from Russian sources on the Ukrainian electricity system – the first ever recorded such instance. Hardly receiving public comment, it could yet become the worst threat of all to national energy security.

Britain's once resilient and liberation power network a weaker and much less reliable system. Here, too, the trilemma challenge has failed to be met and with dangerous consequences.

The Road Not Taken

There are possible ways out of the world energy trilemma but they are not the ones currently being followed. For a start, new technology delivering a far more efficient 'burn' needs to be applied to all fuels, not just to green power sources. Both coal and gas can be burnt with far greater efficiency. Oil and all petroleum products, such as gasoline, can be used with vastly greater efficiency. Electricity can be transmitted with much smaller loss; power can be used with immense gains in savings.

Now supposing, just supposing that policy took a different route and one more in line with current realities. Suppose it was acknowledged, instead of being denied, that coal-burning was bound to continue on a vast worldwide scale and that the only possible way to make this consistent with climate targets was to make the coal-burning process carbon-free. The priority would then become not to penalize coal out of existence as a global fuel but to simplify and reduce dramatically the costs and complexities of technologies to sequestrate and pipe away underground the carbon dioxide emitted. It would then have to be conceded by decarbonizing strategists, however reluctantly, that coal-burning could continue.

CCS at present remains in its infancy and well beyond the reach of most societies, even the richest, let alone the poorer ones who have been bypassed by globalization and are struggling to survive. In effect, if applied to all coal-fired stations today, it would double the cost of coal-fired power which countries like India cannot possibly afford. By contrast, Polish authorities estimate that they can build non-CCS coal plants with 45% greater generating efficiency than previous conventional coal stations – thus substantially reducing carbon emissions per unit of energy output.

Even to get to its present stage of development, CCS has had to draw on very large consumer and taxpayer subsidies, generating inevitable political and public resistance. Over a period of fourteen years, governments have announced commitments of no less than $24 billion for carbon capture and storage projects. But to date there is just one carbon capture system actually in commercial operation – at the Boundary Dam coal plant in Canada. Many have been postponed or scrapped as government funding has been cut or budget limits have been reached and exceeded. But, ever hopefully, more are under construction – three of them at power stations and others at factory plants where coal is being burnt, such as fertilizer producers, possibly about twenty-two in all.

The International Energy Agency says that another staggering $4 *trillion* would be needed to build enough CCS projects to meet the 2050 climate goal deemed so vital in keeping the world's temperature at less than 2°C above 1990 levels. The figure reaches into the realms of absurdity. Besides, while CCS may be seen as the way to carry on burning fossil fuels and yet somehow reach low-carbon goals, it is far from popular for those who want fossil fuels, and coal especially, phased out entirely. So for one camp, the fossil fuel producers and users, it is the silver bullet, although a cripplingly expensive one; for the other, green camp, it is a dangerous distraction from pure green hopes and aims.

One idea put forward is that the resources needed for CCS innovation, bringing its cost down from fantasy levels to practical and commercial ranges, could be spread through time and transferred from the hapless consumer to all the entities burning coal for electricity. That, at least, would smooth the political path and ease the public cost. Another cost-reducing factor might come through the enhancement of existing oil fields output when CO_2

was injected, hopefully producing more revenues to offset CCS expense.

Other ideas for making the whole process far simpler, by processing the carbon on site into other materials, rather than having to pipe it away, may also be round the corner. Better still would be some means of turning CO_2 itself into fuel – the perfect but so far unattainable solution to the whole dilemma. This technology – through carbon capture and *utilization*, rather than carbon capture and *sequestration* – might well prove the more practical one, when and if policymakers face up to it.

Amidst all these hopes, dreams, and fears, the sensible low-carbon path is currently the road *not* being taken. Coal now still provides 39% of worldwide power. That is the fact too many people just cannot bear to face. Banning world coal-burning is an impossibility (and, as we have seen, becoming more not less so) and paying for worldwide CCS, beyond all dreams and all likely resources. The only hard, realistic way forward has to be through more efficient but cheaper and simpler coal-burning methods, *plus* switching from coal to gas wherever possible, *plus* far faster and bigger reductions in the cost of all renewable sources. Of the latter, low-carbon nuclear power must somehow eventually win through, because it is the only one so far that can produce uninterrupted baseload electricity on the scale which modern industry requires (even though many firms are seeking ways of producing their own heavy-load power on site). But as we shall see, before the world can have affordable nuclear power, there are mountains to climb.

Meanwhile, as Indian ministers confirm, and as all evidence from China confirms, discounting occasional public statements for global consumption, the developing world is absolutely determined to put growth and poverty reduction first, which means using the cheapest possible sources of power in growing volumes.

There is one, and only one, practical route out of this dilemma. If coal-burning on this scale is going to occur anyway, then a cheap version of carbon capture has to be devised. Nuclear power alone cannot possibly meet the gargantuan Chinese energy appetite, and coal station construction will continue apace (although the Chinese have now made important steps towards insisting on importing only the cleaner types of coal).

To come anywhere near their new commitment to ensure that carbon emissions 'peak by 2030', the Chinese will also need, on top of many more coal-fired stations, to accelerate still further their current nuclear build. They now have twenty-nine nuclear plants under construction with the aim of generating 58 gigawatts of electricity (that is 58,000 megawatts) by 2020, with, by then, another 30 gigawatts of nuclear power under construction. Even that, say researchers at Tsinghua, will not be enough, and coal-burning will have to increase still further.[2] (See more about this in Chapter Fourteen.)

All estimates which are grounded in reality confirm that fossil fuels – coal, gas, *and* oil – will remain dominant in world energy supply for another century, and probably indefinitely, however ingenious the development of alternatives and however great their fall in cost. Renewable energy flows from hydropower, from wind, and from solar will certainly grow from their present low levels, but they will share the global energy mix with hydrocarbons. Fossil fuels will play a major role in satisfying the world's energy needs for the foreseeable future. Either they are burned cleanly or the carbon targets will be missed by miles. This is the reality that energy and climate policymakers will have to face but which many doggedly deny.

2. *Financial Times*, 14 November 2014.

CHAPTER SEVEN

European Policy: Energy Disunion and the Upside-down Strategy

How EU policy blunders have undermined both energy and climate strategies – Europe miscalculates – all fossil fuels initially opposed – the Russian wake-up call – the Fukushima impact – other avenues blocked – no real European energy market – ambivalence between security and decarbonization.

A rusting coal carrier nudges its way up the Tyne past the glittering, glass structure of the Town Hall. It is carrying Russian coal to Newcastle. A mile-long train of coal trucks trundles across the Lower Saxon plains, bringing brown Czech and Slovak coal (lignite) to Germany's hungry new coal-fired power stations. Vast new coal-mining projects are opening up across Europe – in Brandenburg, on the Scottish Borders, in Spanish Asturia, in the Polish heartlands. Big American,

Australian, Canadian coalers queue to find a berth in the coaling harbours of Northern Europe.

The European vision was different. None of this was supposed to happen. Oil and gas would become impossibly expensive (and peak out). Coal would be taxed and regulated out of existence or use. If EU policy, building on these assumptions, could reinforce the switch to renewables as fast as possible, the EU would emerge as a zone of low-cost green energy and a world leading producer in low-carbon technologies.

This entire structure of beliefs, hopes, and commitments, on which the EU's unilateral policy was grounded, turned out to be disastrously wrong. Instead:

- Fossil fuels stayed cheap – with plentiful coal, vastly increased volumes of cheap (mostly American shale) gas, millions of barrels more oil, and yet all up against subdued world oil demand growth.
- The cost of subsidies soared, a) becoming unaffordable for cash-strapped governments and b) imposing unacceptable energy costs on consumers in Europe. Renewable energy costs fell too slowly, if in some cases at all.
- The Fukushima disaster in Japan scared off nuclear power investment, closing it down completely in Germany.
- Coal trade boomed, with plentiful supplies to Europe diverted from the US and from Australia. China accelerated coal-powered electricity investment. So did India – on a truly massive scale. So did Germany. So did Japan
- Subsidy focus on renewables discouraged gas turbine investment and gas transmission investment, so that the carbon gains from a switch to gas from coal were not secured for Europeans, opening wide the energy cost gap between

the USA and Europe and directly threatening progress with transatlantic trade liberalization (the so-called Trans-Atlantic Trade and Investment Partnership).

- Carbon *consumption per head,* as against carbon *production,* soared in Europe (in the UK alone by 28% between 1990 and 2010). While industrialized nations have been claiming that their carbon emissions have fallen, in reality, emissions embedded in their consumption patterns have soared.
- The major emitting powers took little notice of the EU 'example'. While they recognized the dangers, they argued that the solution lay in the hands of the older industrialized countries. And in the case of India, they reasserted their belief that economic growth came first and that the cheapest energy forms were essential for their progress.
- Attempts to price carbon through a complex system of emissions trading had zero effect.
- Biofuels subsidies pushed up food prices without cutting carbon emissions. Valued landscapes were blighted. Rainforest clearance for palm oil planting accelerated. Many other side effects, especially falling on the poorest communities, occurred.
- In short, the EU vision, born of high ideals, of pursuing a unilateral energy and climate policy which would lead the world, was busted, undermined by events, and shattered by realities. There would have to be a change of direction.

As the previous chapters have noted, in Europe as almost everywhere else, coal-burning is booming. While other power sources are certainly spreading (such as nuclear, wind ,and solar energy) across the globe, the coal trade – import and export – is in massive expansion mode, with coal not only still the world's largest traded commodity and single biggest source of electricity but becoming more so. All this is the diametrical opposite of what the policymakers intended, argued for, planned, legislated for, hoped for.

In Europe, where the government policies aimed at curbing coal have been most elaborate, and the anti-coal rhetoric has been loudest, the return of coal has been fastest and the revival and expansion of coal imports, coal production, and coal-burning most striking. We are looking at a policy fiasco of staggering proportions, bringing more pollution, increased carbon emissions, severely reduced power supply reliability, and a reversal of green ambitions on a major scale – and all at heavy cost as governments and EU authorities struggle ineffectually to subsidize non-fossil fuel alternatives, penalize coal, tax and levy consumers, shut down coal-fired power stations, denounce coal use. All in vain – the opposite has occurred.

Later, we will return to the specific drivers behind this persistent upward global trajectory in coal-burning, strongly reinforced by Chinese industrialization, with Indian modernization trailing a little way behind.

But to understand the particular perversity in Europe – the one place where the hope was to check coal-burning and set a world example in the decarbonization of electricity – we have to go to the heart of both EU energy and climate thinking and policies, and to the heart of the member states' varied concerns and needs.

The EU Imbroglio

The great re-enthronement of coal throughout Europe, when the opposite was supposed to happen, is of course only one aspect of the amalgamation of internal confusions, basic errors, and contradictions at the heart of both European and especially German and British energy and climate policies. Energy prices in the European Union are now amongst the very highest in the developed world. According to the European Commission, medium-sized businesses in the EU pay some 20% more for their electricity than Chinese companies, 65% more than similar firms in India, and twice as much as companies in the USA and Russia, and more than three times as much as similar-sized Middle Eastern industrial consumers, for example in Saudi Arabia or the UAE.[1]

This is an open and official admission that European gas and electricity prices are now doing real damage to competitiveness, as well as, of course, imposing genuine hurt on households. It seems they are due to go higher in the period ahead. Another European Commission paper, published in January 2014, warns that efforts by industry to compensate for their disadvantage (i.e. paying green energy prices far above those of their non-European competitors) 'may need to go even further'. The Commission's authority to push costs upwards in this way derives from the 2009 Lisbon Treaty. Before that, European energy laws lacked any formal legal basis. This did not prevent the EU from developing a rough and ready energy policy built on a series of directives which were justified on environmental grounds, such as the Renewable Energy Directive and the Emissions Trading Scheme.

But after 2009 a cascade of legislative instruments flowed from the newly empowered Directorate General for Energy in Brussels. The outside observer would have assumed that these would be aimed at opening cross-border trade in Europe, the

1. *European Commission Staff Working Document 17,* March 2014, quoted by Matthew Elliott in a Paper for Business for Britain.

mission of the EU being to 'complete a Single Market', seeing that the EU was now specifically entrusted by the Lisbon Treaty with the good functioning and security of energy markets, and their efficiency and interconnection. But the outside observer would have been wrong. The bulk of these directives continued to be driven by environmental concerns and by the cause of Europe-wide decarbonization – the assault on all fossil fuel energy sources by making them much more expensive.

In the British case there can be no doubt, as we shall see in the next chapter, that these European policies were on occasion gold-plated – i.e. applied with additional zeal and ferocity. But the central EU thrust was, and remains, clear. It is to create a European zone of high-cost energy – in practice, one of the highest-cost areas in the developed or fast developing world. This would be done not by encouraging technologies which would allow cleaner and more efficient coal-burning, not by switching energy production from coal to much cleaner, lower-carbon gas, but by making the use and consumption of all fossil fuels so uniquely costly, and investment in fossil fuel power stations, so patently unattractive as against investment in green energy, such as wind farms especially, that a post-fossil fuel future would open up. Europeans would thus escape the dangers of oil and gas which was certain to run out and soar in price before doing so. And there would be a double benefit that, as the rest of the world followed the same route, Europe would be masters of the new situation and, better still, with a head start over the rest of the world in providing the technologies and the equipment that all other countries would be seeking as they fled from oil and built renewable energy facilities.

The perfect transitional fuel

Unfortunately, the rest of the world did not perceive things quite this way. Other nations saw, or their business decision-makers saw, something which the European Commissioners had missed. They spotted that, in reducing carbon emissions, a 1% switch from coal to gas in power generation was the equivalent of an 11% increase in power from renewables. This was the calculation made by Christof Rühl, the former chief economist of BP, which nobody cared to challenge, because it was all too obviously correct. Business leaders also found that the confidently predicted soaring price of gas and the 'peaking' of costly oil did not seem to be happening. A much wiser central thrust of strategy, if it could be engineered, would clearly have been from the start to shift from coal power to gas power – which is of course very much what the Americans have done, albeit under the impulse of market forces rather than government directives, with highly positive results in cutting carbon emissions. But that is not how the EU energy policy planners saw things.

All the signs are that Chinese policymakers would like to go the same way – from coal to gas – and are trying to do so, although it will be a very long haul. The central role of China in the whole nexus of world energy and climate issues will be the focus of Chapter Fourteen. But the basic, startling fact is that 68% of all Chinese primary energy use is from coal and that China's share of world coal demand is now 51% and rising. India, incidentally, relies on coal for 42% of its primary energy, but that too is rising fast.

In Poland also, which starts from a hugely burdensome 90% reliance on coal-burning for its power supplies, the hope is that there can be strong advance along two low-carbon paths – towards more gas production from shale gas and to much more efficient and brand new coal-fired power stations.

Yet here was EU policy with its face turned firmly away from any sort of coal-burning or, initially, gas-burning. The strategic posture was against all fossil fuels and in favour of all heavily subsidized renewables – wind, solar, biomass, the lot.

In consequence, new investment in gas turbines has dwindled to a trickle across Europe, and especially in the UK. Yet to the bewilderment of the authorities and the European Commission, it has led to quite a different outcome from that intended – a major expansion in coal consumption in power stations, plus the construction of new coal stations in many parts of Europe and a rise, as we have seen in the previous chapter, not a fall, in carbon emissions.

'King' coal has returned with a vengeance – in defiance of what was intended – the dirtiest of fuels and the nightmare trend from which the most dedicated green energy campaigners were trying to escape. The low emissions path from coal and oil to gas having been strewn with obstacles, the power companies and their investors have turned to the most profitable and easiest alternatives – not wind farms nor even the now much cheaper solar power sources, but coal, imported coal, indigenous coal, brown coal, open-cast coal, deep coal – coal from any source across the globe.

Major outside world changes have also powerfully contributed to this chapter of mistakes. In the words of Dieter Helm, 'Three external developments finished off the job of undermining the rationale (of European policy).'

The first was the Fukushima disaster, which led to the 'Energiewende' in Germany and the retreat from nuclear power across many European countries, including the usually level-headed Switzerland.

Secondly, the Copenhagen, Durban, Lima, and most recently, Paris world climate conferences put paid to the dream that the world would tamely follow Europe's hopeful lead and sign up to precise, legally binding agreements to reduce carbon emissions in line with agreed international targets.

Finally, the shale gas and oil revolution punctured the belief in the imminence of expensive peak oil and peak gas, and in the process exposed Europe, not the USA, to a massive energy competitive disadvantage.

But there was one further event the EU energy planners had not foreseen – an event so challenging that it began to jolt the basic assumptions on which European energy policy had been built. In April 2012 things began to go very wrong in Ukraine, that country being the chief transit route for piped Russian gas into European markets which supplies some 40% of all Europe's gas needs – in some Central European cases, up to 80%, 90%, and even 100% of daily gas supplies.

There had been threats before in 2006 and 2008 and minor interruptions but somehow these had not deeply stirred the energy planners, who were more concerned at this earlier stage to phase out all fossil fuels. But Putin's moves in Crimea and the outbreak of full rebel hostilities across Ukraine with obvious (although denied) Russian connivance brought a shaft of frightening reality into Brussels thinking. Suddenly, the dogged hostility to gas as yet another carbon-laden fossil fuel began to look badly out of place. A desperate and belated drive to reduce European dependence on Russian gas began to change the tone of the energy planners, with calls for new strategies for European gas supplies, including fracking (after previous deep antipathy), more gas pipeline interconnectors from the western end of Europe, more gas, piped and frozen, from Norway, new routes for piped gas from the Caspian region but not across Russian soil, and more LNG terminals to import frozen gas from a variety of world sources, including the United States.

Policy confusion has been further deepened by two other factors. First, there is the rejection of low-carbon nuclear power in several European countries – notably Germany (a sorry and utterly counterproductive saga of which more in Chapter Eight). Second, there is the strong opposition to shale gas and oil development (fracking) across Europe, both governmental and official, as in France, and popular, as almost everywhere else. In the UK, while official views support shale drilling through deep hydraulic fracturing, and while high-quality deep deposits clearly exist in the North-West, the North-East, and the Midlands of England, domestic shale gas development and production have been largely thwarted (see the 'British fiasco' in the next but one chapter).

Thus, the two most promising energy production routes to decarbonization – gas and nuclear power – have been until recently systematically blocked, and with the burden of crushingly high green energy levies added. Only with the Russian crisis have 'second thoughts' on gas come into the EU policy equation.

There can be no relief from this total – and costly – policy debacle until it is fully understood by EU policymakers why and how this has occurred. As long as gas is banded and bracketed with coal and oil as all part of the hated fossil fuel syndrome, to be replaced somehow by renewable energy regardless of cost, Europe is going to be denied the one path to lower emissions which really works and which the US so obviously enjoys, namely a big transitional switch from high-emissions coal to lower-emissions gas for electricity generation. Instead, cheap coal is going to stay on top in the drive for new generating capacity in an electricity-hungry era.

Whether the Russian jolt succeeds in waking the policy dreamers or moves them on from mere rhetoric and paperwork remains to be seen. Indeed, it is less a matter of dragging the policymakers out of a reverie than of demanding a complete change of direction. Is this even faintly likely?

The question has to be posed because we are dealing with deep-set beliefs, upset by events, rather than carefully evolving policy formulations. From a cavalier disregard for the gas sector, and intense discouragement to invest in new gas plants, or indeed keep existing ones open, the European policy emphasis has jerkily twisted into building up gas capacity at all costs. The driving assumption takes over that the Russians will actually cut off gas supplies not just to bankrupt and disrupted Ukraine but to its main European customers – a third of its total export market. This assumes total irrationality on the Russians' part – always possible if we remember Winston Churchill's depiction of Russia as a riddle wrapped in a mystery inside an enigma, but somehow unlikely.

But just the opposite is also possible – namely that Russia will make intense efforts to sell MORE gas at much MORE competitive prices to Europe. Indeed, there are strong signs that it is already doing so – despite abandoning its South Stream gas pipeline project under the Black Sea, in the face of strong European Commission opposition.

The European policymakers, like the crew of a drifting and anchorless ship, could thus find themselves swinging from one catastrophe to another – from costly and perverse opposition to gas, to blind support for alternative gas sources, again at debilitating cost. From being unable to envisage gas as one of key carbon-lowering and abundant fuels of the future, they could now be in danger of embracing gas at any price so long as it does not come from Russia.

The Market that is not there

Almost everything, explains the European Commission, which it does in the field of energy policy is 'designed to complete the Single Market, to extend it to the field of energy, and to render it more attractive for suppliers and consumers'.

Such is the state of mind of the well-intentioned officials in the Energy Directorate at the European Commission in Brussels. Such is their explanation, their strategy, their goal. Yet to this pinnacle, this central aspiration, there is a delusional element. A single market in energy supply would exist if the physical infrastructure to convey electricity by grid or gas by pipeline to all corners of the market and all member states in fact existed. But it does not yet do so, however strong the aspiration. This means that the string of competition rules aimed at the European 'energy market' can certainly have an impact on parts of the supply system, but they cannot order the whole European market because that market is not there. Gas and electricity cannot be bought and sold freely in large parts of the Union. This Single Market emperor has no clothes, although few dare say it.

The situation is akin to the Eurozone system, where again a massive roof has been created without walls to support it. To work properly, the Eurosystem requires political union, or at least a degree of budgetary union which would involve intimate political involvement by the EU authorities in the inner financial affairs of all member states. But nothing like the political union is there. Member states are simply not ready to concede that degree of sovereignty. So the Eurozone, like the structure of regulations and rules governing energy supply, exists, as it were, in mid-air, unattached to reality on the ground.

In the case of the Euro, the precarious and distorting results of such a bizarre architecture have long since become clear for all to see. Constant makeshift propping-up attempts have been necessary to meet crisis after crisis. In the case of EU energy market rules and energy policy, the impact of building castles in the air has led to similar but less visible distortions.

Meanwhile, the catalogue of central energy questions facing Europe remains unanswered.

Now that policy has veered from gas hostility to gas concern, can Europe wean itself, at least to some extent, off Russian gas? Can the EU restore the competitiveness which high energy prices have drained away? Can the intense German energy predicament, largely self-imposed but damaging all Europe, be disentangled? Can the UK energy muddle, raising prospects of actual power cuts, be unravelled in time? Can France retain its nuclear (and low-carbon) pre-eminence? Can new pipelines and interconnectors bring fresh energy supplies to Europe and create a genuine energy market within Europe?

Current European policy, knocked off course and rudderless, answers none of these questions. Can it even clarify whether the previous policy or the new Russian-shaken one has actually contributed to a significant reduction in greenhouse gases or to the global battle to curb carbon emissions? Are there ways in which the European powers, either individually or collectively, can begin to lift themselves out of the mire of insecurity and dangerously high costs into which their energy policy has been allowed to drift? How might a common energy policy for the whole European region begin to emerge?

Very recent proposals from the European Commission for a EU Energy Union make a brave attempt to do just this, as in Chapter Ten we shall also attempt ourselves although taking a different route. We shall come to some tentative answers but to do so we need to drill deeper into the complexities and confusions as they are working out in individual national situations.

First, we turn to the Russian enigma, then to the German muddles, then to the Polish fury and discomfort with EU energy policy.

After that, in the next chapter, the focus will have to be on the most embarrassing (for Brits) and awkward question of all – how does Britain, home of the Industrial Revolution, mistress of steam and power for two centuries, highly successful pioneer of North Sea development (Europe's major oil field), escape from the total mess and inadequacy into which it has been pushed by its misshapen energy and climate policies? How can a nation of this maturity now face power interruptions with talk of blackouts?

CHAPTER EIGHT

Russian Dolls, German Muddles, Polish Dismay

What does Russia really want? – Central Europe's Russian gas dependence – the Crimea-Ukraine 'shock' – other European gas sources considered – Russia looks east.

The German energy policy muddle – Energiewende and the massive expansion of the dirtiest kinds of coal-burning – damage to German industry.

Polish dependence on coal – shale hopes dashed – Polish clean coal ambitions – hostility to EU renewables plans – an example of EU rigidity.

What are the Russians up to? Do they want Ukraine back? Or do they want to carve out a new republic of Russian-speaking areas, backing violent methods to do so? Or do they want somehow to regain parts of the old Soviet Empire?

Nobody knows, but whichever way Vladimir Putin is seeking to take modern Russia, there are heavy energy implications for Europe and strategically important judgements and decisions to be made.

First, the facts. As mentioned, Russia supplies about 38% of EU gas imports, roughly half of which is currently piped through Ukraine. Russia was until recently receiving a regular payment of $5 billion per month for its EU gas exports. But the overall figure hides the real story. For the ex-satellites the percentages are far higher – in the case of little Estonia, 100%, and Estonia is still also linked in to the Russian electricity grid, in the cases of Bulgaria, 90%, Hungary, 80%, Slovakia, Romania, the other Baltic states, all about the same. Germany itself takes 22% of its colossal daily gas needs from Russian sources – of course, a far higher volume than any of its eastern neighbours' needs.

Not surprisingly, Germany has focused in the past on bilateral deals with Russia, with European solidarity coming a poor second. Gazprom has signed agreements in recent years with BASF and E.ON Ruhrgas. Poland takes 70% of its gas from Russia, although gas is only a small part of its largely coal-based energy system. The Poles are trying hard to reduce that figure. They are now able to take LNG from Lithuania's big new floating LNG facility, now connected up. Italy takes 15%. But for the rest of Western Europe, with plenty of their own gas sources, the percentages are negligible.

Overall, the volumes are big enough to create a wide sense of dependency and, of course, a sense of vulnerability should there be interruptions for whatever reason.

And there have been

Both in 2006 and again in 2008, Gazprom, the state-directed gas supplying giant, cut off supplies of gas to Ukraine over payments disputes. Knock-on effects on other European customers were immediate, since the Ukrainian authorities did the obvious thing and tapped supplies destined for Central and Western European customers to meet the shortfall. However, as we have learnt, these warning shocks were not big enough to stir European Union reactions at official or Commission level.

That had to wait for the much bigger shock of 2013 when Russia annexed the Crimea and made evident its longer-term intentions to interfere in internal Ukrainian affairs and actively promote violent separatism.

The issue of dependency on Russian gas then came to the fore in a prominent and almost emotional way. Suddenly, the need to find alternative gas sources burst to the top of the EU energy agenda, displacing earlier strategic priorities which had placed decarbonization and the downgrading of *all* fossil fuels at the centre of EU energy policy.

Now, all of a sudden, the talk was of building new LNG importing plants on European coasts, of importing American gas (hopefully much cheaper but in fact, after transport costs, the same sort of $8-$10 price), of more reliance on Norwegian gas and of the development of shale gas deposits believed to be considerable throughout Europe but not hitherto taken very seriously. In fact, France had imposed a prohibition on all shale gas drilling while, as we shall see in the next chapter, the UK had seriously bungled its plans for accelerating shale gas drilling and exploitation of its very substantial reserves in North-West and North-East England.

The new machismo stance against Russian gas was further boosted by European Commission determination to frustrate new Russian plans to build a major gas supply line under the Black Sea, named South Stream, bypassing the awkward Ukrainian problem and bringing welcome supplies to Bulgaria, Romania, and South-East Europe generally. Bulgaria had happily signed up to the Russian project but was now 'instructed' by Brussels to drop its collaborative stance with Russia. This pressure succeeded. Visiting Ankara in early December 2014, Vladimir Putin announced that the Russian South Stream project had been abandoned. He argued that the Europeans were the losers (in the case of Bulgaria and Romania, they probably were), and that new pipeline plans would be drawn to supply Turkey – and through Turkey into South-East and Southern Europe.[1]

In practice the Russian dependency issue is best looked at from a less emotional and more down-to-earth level.

At the practical level, the problem comes down to four or five Central European countries which are unquestionably over-dependent on one supplier – Gazprom – and therefore vulnerable not just to interruptions but also to brutal monopoly price pressures and to divide-and-rule sales tactics by the giant Russian supplier. Common sense would anyway dictate that any country should avoid too much reliance on one supplier, and situations such as in Estonia, where 100% of gas imports from Russia were combined with reliance on the Russian electricity grid as well, were clearly unwise and had been all along.

Steps to ease this kind of situation by ensuring the construction of more gas pipe interconnectors between plentifully supplied Western Europe and its more vulnerable eastern partners

1. Russia is now enthusiastically booking another new pipeline project, Nordstream 2, connecting the major gas fields of Western Siberia to Northern Europe.

made, and continue to make, complete sense. In fact, they are a necessary part of creating a real market in energy supplies which at present is largely fictional and exists only in the heads of regulation zealots in the European Commission.

A second common sense reaction would be to unblock the many difficulties and policy obstructions in the way of piping more gas into Southern Europe from the Caspian region via Turkey. The complex story of the ill-conceived and failed Nabucco gas pipeline project via Turkey comes in a later part of our story.

But the largest common sense stance of all, overshadowing all these ameliorating possibilities, would be to recognize a central fact about Russian gas supplies. They will continue, they will be large, they will be reasonably reliable, and the price can be restrained by perfectly sensible moves amongst the main Gazprom's European customers to get their purchasing together and not be taken to the cleaners as Gazprom picks off individual markets and fixes the prices at will.

Russia may continue being a big supplier but it is bound to be the overall loser in these developments. Its powerful monopoly position will never look so good again. It will have to face price competition in Europe as never before. Some commentators have even suggested that by 2020 Central and Western Europe will be able to do without Russian gas altogether, arguing that cheap and plentiful LNG will be rolling in from a gas-rich United States to a growing number of receiving points on the European continent.

But this kind of extrapolation needs to be viewed with care. It makes sense to break the current Russian monopoly, but if pipeline gas from Russia comes down in price, it makes no sense to go over completely to imported LNG, which could well be much more expensive.

One day – and it may be sooner than feared – Russia will be freed of Mr. Putin and his uncooperative and anti-Western stance. His interventionist policy in Ukraine has already cost Russia dear, and slumping oil prices have added significantly to Russia's economic difficulties. His newly stepped up involvement in Syria, wheeling in planes and heavy weaponry to rescue Bashar Al-Assad and the Syrian Government, could have further repercussions, both bad and good. The bad we may have already seen in the horrific destruction of a Russian holiday flight over Sinai, which may well have been the bomb work of ISIL (Daesh). The good could yet come from a successful channeling of the Putin initiative into concentration on the poisonous ISIL, rather than Assad's foes generally.

But any moment, his apparent popularity with the Russian people could well turn to disdain or worse when shortages continue, food becomes scarce, and anger breaks out onto the streets.

Investment in new LNG handling plants, like most energy infrastructure, is capital-intensive and needs a long life of steady returns. Political life, even in enigmatic and dictatorial Russia, can be very short-term and volatile. Some new LNG-importing facilities will be useful. For example, Lithuania's new floating LNG import facility is working well. But too many will be expensive white elephants.

There is a further factor winding through the endless intricacies of the energy system. Although it has so far not been the case, one outcome of a prolonged period of lower oil prices could be to weaken US investment in new shale developments and gradually reduce the prospects of plentiful cheap American gas for export to the rest of the world. A more certain effect of collapsed oil prices is to put a stopper on shale gas development round the rest of the world, where costs may well be higher than in America, and where the basic infrastructure and conditions that make the American shale story so dynamic and resilient, do not exist.

The price will of course pick up again one day (see Chapter Eleven) – all commodity prices do. But the very possibility of lower crude prices once again altering the whole global pattern of oil and gas supplies, even temporarily, is a strong advisory warning to gas importers to keep lines open to all sources, with maximum freedom to switch when events and geopolitical occurrences take place.

The new conundrum for gas planners at the EU level is where exactly to turn, and at what cost. New LNG-importing plants could be dotted round the European mainland continent at large expense, only to find themselves uneconomic as against piped gas from much nearer by. The Central Asian gas supply scene could open up and prove valuable to European security, or some of that gas could instead go eastwards to thirsty China and elsewhere. A new southern corridor could bring Caspian gas from the Shah Deniz consortium in the Caspian Sea through Turkey to South-East Europe and Italy. And both Cyprus and Israel have both made large gas discoveries (with the Israelis now producing and exporting gas to Egypt).

All this could go right, or it could go very wrong. Political risks are everywhere and the eastward pull is growing stronger all the time. Chinese companies have already taken a substantial upstream interest in Turkmen gas fields and in the construction of eastward-flowing pipelines. A new pipeline from Turkmenistan could also supply gas to markets in India and Pakistan, maybe tomorrow's best and fastest growing markets.

The realization in Moscow that the European market is becoming less easy has predictably turned Russia towards new Asian customers. Top of the list is China with whom Russia has long been haggling about price. The deal announced with a flourish in October 2014 for Russia to supply 400 billion thermal units of gas over twenty years to the Chinese market was vague about price and looked suspiciously as though it was all about political timing. Russia needed to show Europe that it could find customers. When the gas will flow, the questions of where it will come from (probably new fields yet to be developed), and what actual price the Chinese will pay all remain up in the air.

The German Muddle: Deutschland Unter Alles

Take the European energy story back to one evening in the last century at the German Embassy in Belgrave Square in London. The year is 1980. A dinner is being given in honour of the British artist Henry Moore, a much admired figure in the then Bundesrepublik. About thirty people are gathered round the long table at the far end of the vast drawing room on the first floor. Both the German Chancellor, Helmut Schmidt, and the British Prime Minister, Margaret Thatcher, are present.

The conversation is flowing gently but suddenly there is a pause. Mrs. Thatcher puts down her knife and fork with a crash. 'Did I hear you say, Helmut, that West Germany is planning to rely on the Soviet Union for 15% or more of its daily gas supplies? In that case Germany has gone mad.'

Helmut Schmidt is un-phased. 'But my dear Margaret', he replies – always a little bit patronizing with the British leader – 'you don't understand. The Communist bureaucrats are always the most reliable. They never break their contracts. They always supply as agreed. They are utterly reliable.'

Thatcher was deeply unconvinced. Whether she could see further ahead than the German Chancellor one can only guess, but within quarter of a century, a non-Communist Russia was playing a very different game. Schmidt's reliable Communism had gone, with Vladimir Putin's resentment against the West and aggression in Ukraine in its place. Dependence on Russian gas supplies, not only by Germany, but especially by the former Soviet satellite states, had grown massively, but so had Russian unreliability.

That Belgrave Square dinner all those years ago was the beginning. From that evening there grew what a *Financial Times* editorial thirty-five years later in November 2014 came to describe as 'the costly muddle of German energy policy'.

German business leaders have been the most outspoken, claiming that their country's energy policy, the so-called 'Energiewende', 'will kill German industry'. But they are not alone. Power analysts worry about blackouts; voters, households, and everyday consumers are furious about ever-rising fuel bills; and environmentalists look with dismay on a scene which was meant to be becoming greener, but is now being browned and blackened with coal on a greater scale than for many years past.

In 2014 brown coal, the most carbon-intensive of all primary fuels (as well as being the most directly air-polluting), rose to providing 26% of Germany's electricity supply. More coal stations are being built, more coal mines, planned. Where burning coal does not fill the gap, more is being piped in from Russia and other shortfalls, made up from importing French nuclear-generated electricity.

How could Europe's industrial giant, the leader and motor of the continent's economy, have brought itself to this extraordinary situation of chaos and reversal from all its proclaimed aims and goals? And is there now a way out?

The German green story began to take off in the early 2000s. A renewable energy law was pushed through the Bundestag promising twenty years of guaranteed prices and priority access to the power grid for anyone installing wind pylons, solar power, or other renewable sources.

These bountiful guarantees were to be paid for by heavy surcharges on all electricity bills, although from the start the most heavy energy-intensive sectors, such as chemicals, demanded and received relief from the full extra imposts.

This was meant to create a predictable framework for future energy investment on the assumption that fossil fuels were not only adding dangerously to carbon emissions but were also set to get much more expensive and less reliable. It all seemed to make sense.

Almost from the start, problems set in. First, green energy sources in Germany tend to be in the north, with major offshore wind farms in the North Sea and Baltic Sea. But maximum energy hunger is in the south, especially around Stuttgart. A major new system of transmission lines needs to be built, but this has met popular resistance at every turn. In consequence, the lines have not been built or are way behind schedule, and the power cannot be delivered.

Even if it could, a second problem is that it would be, like all wind power, highly intermittent and requiring fossil fuel backup to even out supply. But because so much highly subsidized green energy comes into the grid, even if it cannot be properly distributed, the spot price for German electricity is far too low to encourage investment in new gas turbines for backup. That role therefore falls to brown coal power stations with their much lower costs.

Then has come the body blow to the whole German energy balance – the policy of U-turns on nuclear power, ending in the decision by Mrs. Merkel, following the Daichi Fukushima disaster in Japan, to switch off eight reactors immediately. The rest are to be closed by 2022.

The outcome has been a goulash of distortions and anomalies. The renewables surge has brought riches to green energy suppliers as they enjoy the huge subsidies. Investors in, and installers of, solar power have done exceptionally well (although not the actual European manufacturers of photovoltaic cells who have been undercut by much cheaper Chinese suppliers). Bavaria now has more photovoltaic units installed than the whole of America.

On sunny days excess solar power goes into the grid at a loss, and as it attracts a fixed price, the subsidized gap between this and the spot price widens and the cost to consumers increases. When the sky clouds over, brown coal stations cut in, adding ferociously to both short-term pollution and longer-term CO_2 emissions.

More and more German industries resort to their own on-site power sources for meeting the high price and avoiding the cost and volatility of outside supplies. The funds for meeting the high prices promised to renewables operators are therefore curtailed and more charges are piled onto the remaining users on the grid.

Trapped on all sides, new German projects and the investors behind them are looking to cheaper energy regions to locate – the US with its wondrously cheap gas in particular. Thus, German energy policy is beginning to saddle the whole massive German economy. Other perversities abound. More Russian gas has had to be imported; French nuclear electricity has had to come to the rescue.

But the haphazard German policy not only penalizes domestic markets. Because of Germany's size and centrality as the main engine of European economic growth, the muddle spreads out to damage the whole European scene. The policies of Mrs. Merkel and her advisers have directly weakened and destabilized

Europe energy security. The nation that has been at the forefront in building greater European unity is striking directly at the heart of EU prosperity and the sustainability of the EU model.

And now Poland

To get a rounded picture of Europe's energy confusions, the Polish example adds a further insight.

The Polish nation lies deep in the sympathies of Western Europeans, perhaps especially the British. It has been a country of endless tragedies and endless courage. Its emergence as one of the front-rank nations of the twenty-first-century European Union is one of the Union's clear achievements to balance against the many less happy setbacks. It would be the natural expectation that everything possible would be done by the lords of the EU to embrace with pride such an ancient nation and one so eager to play its full part, after years of suffering, in the new Europe.

Not regrettably so.

The whole Polish experience, as it grapples with EU energy priorities, tells quite a different story.

Dominant coal – again

Some facts.

Poland's most abundant energy source is coal. Cheap and plentiful coal provides 58% of the country's primary energy mix. This share, however, has fallen from 76% in 1990. The share of natural gas, however, *has* increased from 9% to 13% over the same period. This share remains much lower than the EU average (24%) but will change if investment in shale gas ever takes off.

Surprisingly, because of its own immense resource of domestic coal, Poland has one of the lowest needs to import energy among the EU countries. In its 2010 report, *Polish Energy Policy until 2030*, the Government expressed its commitment to reduce its dependence on fuel imports and diversify its energy sources especially for power generation. Poland is hoping for its own shale gas revolution. The Government has proposed a new law to foster investment in this sector. The revolution, if it materializes, will mean *less* dependence on Russian gas for both Poland and Europe.

Poland imports around 86% of its gas from Russia. Energy security has been top of the Polish Government agenda since the 2006 Ukrainian-Russian gas price dispute. Although short-lived, Russia's decision to cut off all gas supplies passing through Ukrainian territory in 2003 and again in 2009 led to widespread disruptions across Central and Eastern Europe.

It was therefore not surprising to hear the then-Polish Prime Minister, Donald Tusk, describing shale gas as Poland's 'great chance'. But not all is going well. Global petrochemical countries which flocked to Poland in search of shale gas are turning away in disappointment. Exploratory wells have failed to deliver and regulations have added frustrating delays.

Meanwhile, more than 90% of power generated in Poland remains based on coal. This in turn provides Poland with a cheap power base on which to develop its competitive edge and build Polish economic progress. But this is what the EU does not like. It sees the situation as directly undermining the EU emissions reduction target for 2030 of 40% as compared with 1990. It wants Poland to produce – it is the world's ninth largest coal producer – and burn much less coal.

There are four ways Poland can respond to this pressure.

The most risky and expensive is to import much more gas which would come mostly from Russia. A second escape route would be to import much more LNG from non-Russian sources. An LNG terminal is planned in Swinoujscie on the Baltic coast. The terminal's capacity is expected to be 175.5 trillion cubic feet (with possible extension up to 265 trillion cubic feet). A supply contract was signed with Qatar which will provide 53 trillion cubic feet of LNG starting from 2015.

A third way – also risky – is to build hopes on shale gas through fracking – which looks decreasingly likely in Poland.

The fourth course, and much the most sensible, is for Poland to use its massive coal reserves but to build much more efficient coal plants, using the latest technology such as super-critical boilers – and achieving up to a 45% increase in output per tonne of coal burnt and thus significant reductions in CO_2.

But this is what EU policymakers are against, and doggedly remained against even while formulating their revised targets for emission reductions for 2030 – with which Poland has understandably declined to comply, and even while, in another part of the woods, the EU is seeking ways of reducing the continent's dependence on Russian gas.

We have here in the Polish situation an example of the Alice-in-Wonderland cat's cradle of confused and contradictory initiatives to which EU energy policy has been reduced, with climate and security initiatives clashing with each other and working against, rather than in favour of, competition and jobs. Some hope for a gradual return to energy sanity in Europe must lie in the appointment of Donald Tusk, former Polish Prime Minister, as President of the European Council. Polish energy and climate dilemmas effectively highlight the contradictions and flaws in EU policy.

Even more recently, Polish politics has started to catch up with energy realities. The new Law and Justice Party Government under President Jarosław Kaczynski has begun talking about vetoing EU decarbonization plans completely.

It is no surprise that Poland has all along spearheaded opposition to the EU's one-size-fits-all mandatory targets for reducing greenhouse gas emissions, although it has been also joined by Hungary, Romania, and Bulgaria. A further twist to the Polish story comes from the plan for a Trans-Atlantic Trade and Investment Partnership (TTIP), still currently being argued over. The Poles argue that American goods, manufactured with cheap energy, already have a big edge over European products. TTIP would allow still freer European entry. If energy costs in Europe are now to be lifted further by still higher charges for CO_2 emissions, the American advantage will grow much larger. Per capita carbon emissions in the US are still over twice those in Europe (16.5 tonnes per capita as against 7.35 in Europe). Yet they are un-penalized. The cost of a decrease per tonne to EU mandatory levels would be $600 per tonne. It is easy to see who would be the winner between the two sides of the Atlantic, even with present EU policies, let alone with threatened higher carbon charges and costs.

But to understand the full state of perversity into which EU energy policy has tumbled, and the dangerous futility of attempting to standardize and harmonize all aspects of policy across an entire continent, we have to cross the Channel northwards to the nation which was once the heartland and pioneer of energy revolution and industrial revolution and is now the realm of energy chaos – Albion, Great Britain, the United Kingdom (just).

CHAPTER NINE

The Coalition Legacy: A Very British Fiasco

Frailty of the UK energy system – the total trilemma policy failure – the blame game – a case for the defence – National Grid and keeping the lights on – coal dismissed but coal persists – picking the renewables winners – the capacity payments muddle – nuclear renaissance and the Hinkley C problem – a change of direction begins – the constraints of EU policy.

The previous chapter has shown how the European energy policymakers have become diverted onto the wrong tracks with highly damaging results, how the Russian gas issue has added further upsets and misapprehensions, how German energy policy has lost its way, and how Poland has had enough.

But there is one place where these quandaries have all been exacerbated and that is the United Kingdom. There can be no doubt that the erratic course of EU policy, with its changes of direction, its invalidated assumptions, and its belated attempts at adjustment, has inflicted severe damage on the British energy scene and on British industry generally.

But add this to home-grown policy and we have almost the ultimate in bad outcomes and confused measures, culminating in the failure of the final test of all energy policy – its capacity to deliver secure and reliable supplies at affordable prices to the whole of modern society, to its vital industries and services, to its homes, and to its arteries and nerve systems of communication and cooperation upon which its very existence depends. Instead, the British have seen their key guardian of the electricity system's resilience and the main transmission authority, National Grid, struggling with expensive makeshift arrangements, to meet the expected load.

It seems incredible. Great Britain was the cradle of the Industrial Revolution. Good fortune and colossal enterprise gave Britain the oil- and gas-rich North Sea. Yet this is the same country which now faces power shortages and blackouts, with record high power prices adding injury to the uncertainty. As the spare cushion of capacity in the electricity-generating system drops to danger levels, the media rush to reveal what has been plain for a decade past – that policy incompetence on a grand scale has failed to prepare or provide for a dangerously fragile power supply scene, and in the years ahead may well continue to fail to do so.

This fragility has emerged despite there being plenty of basic fuel supplies – indeed, no shortage at all in the primary fuel supply system. It is true that the North Sea province itself is plainly now declining from its glory days – as oil fields do – and Sir Ian Wood has warned that 'the UK should prepare for a future without oil'. But the country is surrounded by gas producers eager to pipe more to the British market, and the high seas are loaded with cargoes of oil and frozen gas from a variety of suppliers looking for a buyer.

A scarcity and shortage of supplies is not the problem. Nor is soaring demand for power causing the strain. On the contrary, the overall demand for electricity in Britain has been falling during the recession years. As the economy goes into faltering recovery, demand picks up a little but there have been major gains in efficiency. The consumer at home or in industry has done his or her bit. Demand as of 2015 is just about flat.

The problem lies elsewhere. It lies in the failure of policymakers to provide the right conditions in which investors are prepared to risk their funds in financing new generating capacity – or put another way round – the problem lies in the actions of policymakers who have created precisely the wrong conditions for new investment.

Why is replacement capacity so necessary? Because older coal-fired stations have all been put on a closure path by EU directives,[1] because the British nuclear power fleet is old and near the end of its life in almost but not quite all cases, because past decisions to authorize or encourage new power station building have been ducked and delayed again and again until almost too late.

Now National Grid has been pushed into the limelight and forced to make a series of statements about emergency measures.[2] Older coal stations that were due for closure have been put on standby; mothballed gas turbines have been incentivized (i.e. paid) to start operations again with the cost being charged to consumers. The old nuclear power workhorses have been given a few more years by an indulgent safety authority. The giant and ancient Dungeness B nuclear station has been given a ten-year lease of life extension (to 2028) by the British Office of Nuclear Regulation,

1. EU's Large Combustion Directive (2001/80/EC).

2. In July 2015 National Grid warned that the safety cushion of spare generating capacity could fall in the next winter to 1.5% of the available total. The previous year it had been 4.5% and four years ago, a comfortable 17%.

and at a cost of £150 million in extra investment in the plant – a tiny sum compared with the colossal figures for building new nuclear stations.

Meanwhile, on the demand side, industrial consumers have been warned about interruptions and offered so-called 'interruptible contracts' (supply contracts which are cheaper on the understanding that there may be interruptions). Customers would be well advised to install their own emergency supply plants. Lower voltages for households, meaning dimmer lights and it taking longer to boil the kettle, may well be on the menu.

As for household consumers, a pitiful scene has begun to unfold. Fuel banks have sprung up across the country to help desperate homeowners forced to cut off electricity for which they simply cannot pay.

Research by the Citizens' Advice Bureau suggests that as many as 1.6 million UK families are actually disconnecting their meters in desperation as they face the brutal choice between food and warmth. The suffering behind these figures constitutes a major national disgrace. Minuscule sums are being distributed in limited areas of the country, just enough to restore a small intake of current and bring a little home warmth. Ironically, the figures of sums distributed come in at less than the costs added to all energy bills by green subsidies.[3]

3. Official UK Government estimates claim that green levies and charges in various forms in 2014 cost a typical household £68 a year or 5% of an annual energy bill of £1,369. They are forecast to rise to £141 by 2020, and £226 –15% of an average bill – by 2030. The Institute of Public Policy Research in a July 2015 report argued that these levies were regressive, adding that there had been 'insufficient debate about the dramatic rise in charges on energy bills that is set to take place over the next decade to fund progress on low-carbon and social policies'.

Overall, National Grid, a company of consummate skill and experience, believes that one way or another it can keep power flowing in any emergency for the next three winters. When the wind ceases to blow across Northern Europe, or when a power station goes down through some mishap or for technical reasons, Grid managers believe that – at a cost – power cuts can be avoided. Extra generating capacity of many different kinds can be paid to be on standby. Extra inducements to big power consumers can be offered for them to cut their consumption at peak times, or switch to their own generators.

Beyond 2018, we shall see, is another matter. There it becomes a question of somehow inducing energy and power companies to invest in new generating capacity that can be kept on standby for years ahead to swing into operation when the growing supplies of wind power cut out, as happens quite often. We shall see how attempts by the British Government to fill this future ominous gap have led to bizarre results and incredibly complex and costly patch-and-mend measures.

This is a scenario which might better belong to some of the world's poorest countries, although with new power technologies leaping ahead, such as cheap solar power, it is one which they may be able to avoid. The so-called advanced economies like Britain suddenly find that they are going back, not advancing – at the rear, not the front, of the race.

The Blame Game

There is a certain unfairness in always blaming ministers personally for policy failures in their departmental areas. Departments cover vast areas. No single mind could possibly cover the full range and variety of departmental activities. Things happen. Government actions are anyway supposed to be collective. Of course, ministers are held accountable. Heads are demanded and ministers fall. But sometimes it can be the roughest of justice.

No such scruples need apply when it comes to the two most recent Secretaries of State for Energy. These two, Chris Huhne and Ed Davey, were successively the ninth and tenth holders of this post since it was created in 1974 by the then-Prime Minister Edward Heath.

Their predecessors had been Peter Carington, John Varley, Tony Benn, David Howell (author of this book), Nigel Lawson, Peter Walker, Cecil Parkinson, and John Wakeham. In the Blair years the post was downgraded and the department, merged in with Industry for a while, until an awkward marriage with climate change issues, overlapping with environment matters, brought it back into being as the Department of Energy and Climate Change, with Ed Miliband as the new overlord.

No particular glory or personal stamp attaches to, or was ever claimed for, the reign of the previous incumbents. There were triumphs and failures. Lord Carrington struggled with oil shocks and miners' strikes. The late Tony Benn deserved credit for getting the North Sea oil programme going. Lawson completed Howell's work on oil industry privatization and preparation for further inevitable challenges from militant miners. Walker handled the next miners' strike, when it came, with aplomb. Parkinson and Wakeham carried through further complex energy industry privatizations, and much else was achieved in modernizing the whole energy sector all along the way.

But none placed themselves so prominently at the head of total policy reversal and energy revolution as the three most recent incumbents, three zealots, one inside a Labour Government, two in the succeeding Coalition. And none failed so spectacularly and personally to either achieve their declared aims or prevent severe setbacks and damage to the whole energy sector.

Here it has been not just a question of accountability, or media victimization, or nobly shouldering responsibility for what has gone wrong, but active and precise association, in the face of numerous warnings, with all the measures that have led to such disastrous outcomes for the British energy scene.

Consider the following.

First, from starting with an ample margin of surplus electricity-generating capacity, fully adequate to meet the highest of peak demands, to cover emergencies such as major storms or breakdowns in any parts of the system – running for many years past at between 12 and 20 gigawatts of electric power – the margin or cushion in 2016 was allowed to shrink to barely 2 gigawatts and maybe to nothing. The system has become dangerously vulnerable, even more so than would have been the case in the past because of the heavy volume of wind power on the system which can shut off suddenly at any time, depending upon the weather.

Second, delivered energy costs, both to industry and to the householder, have soared to unparalleled levels, some of the highest in Europe, up by 40% in real terms since 2010.

Third, and in consequence, fuel poverty, by which is meant the number of households facing energy bills of more than 10% of total disposable income, has risen to a record level of seven million families, with consequences as described above.

Fourth, energy-intensive industries, while struggling to cut energy use and receiving a degree of special assistance, have nevertheless seen markets turn away to cheaper sources with consequent major job losses. Large closures throughout the British steel industry, beginning with Redcar complex, exemplify the energy cost problem. A carbon tax unique to Britain has shifted British business costs still further out of line with competitors.

Fifth, the most carbon-intensive form of electricity generation, burning coal at power stations, mostly imported from Russia, America, and Australia, has risen as a proportion of total electricity supply to the highest levels since 1990.

Sixth, while carbon emissions per unit of production have fallen during a five-year period of deep recession, overall carbon consumption, taking into account the carbon emitted in the manufacture of Britain's substantial imports, has risen sharply – by about 20% since 2010.

As described in an earlier chapter, the three goals set by energy policymakers, led first by Labour's Ed Miliband and then by the two Coalition Secretaries of State, have failed spectacularly to be reached. Affordability has been disregarded, reliability has been minimized, and overall decarbonization targets are being missed (when emissions are fully measured). Intense environmental damage has been done to the whole of Britain's countryside by wind pylons, with not even the most precious beauty spots spared, and a heavy redistribution of wealth has seen cash from the poorest households funnelled on a major scale into the pockets of lucky landowners and wind power companies.

How could such appalling damage and such hostile and counterproductive policies come to be inflicted on the British economy?

Like Mr. Miliband before them, the two post-2010 Secretaries of State were each gripped by two powerful convictions. These were, and are, convictions which, if held with balance and pursued with moderation, command widespread respect, but if pursued and applied to extreme degrees to the exclusion of other influences, can turn lethal. Like many common medicines, if taken carefully they prove beneficial, but if taken immoderately, and even more if combined with other drugs, they become completely toxic.

The two convictions held by Mr. Huhne and Mr. Davey were:

First, that fossil fuels must be phased out of power production, and if possible out of all energy use entirely, in the name of combatting climate change; and second, that Britain must at all costs adhere to the laws, aims, and direction of the European Union (as related in the previous chapter), as well as to its own uniquely constricting Climate Change Act of 2008. Indeed, Britain should go beyond EU law in its zeal to cut carbon emissions.

Putting these two beliefs together, combining absolute determination to conform to the direst and most precise predictions of climate scientists about carbon emissions from fossil fuels with equal determination to abide by all EU targets for reducing emissions Europe-wide, all EU directives for phasing out coal-fired power stations and all other related EU rules and regulations, has created the perfect storm of bad outcomes for the energy situation and bad prospects for the whole United Kingdom.

This combination of EU energy policy with home-grown zeal has produced a deadly brew for British industry The official estimate for 2015 was that EU energy regulation and targeting requirements would impose a net cost on British industry of £86 billion a year. The two major EU requirements – the Renewables Obligation and participation in the Emissions Trading Scheme – are estimated to add 9% to energy costs. Opting out of the renewables target (the obligation on electricity companies to purchase a fixed percentage of renewable electricity at heavily inflated prices) would reduce energy costs by 7%.

The EU policy of high energy prices, combined with the cost of all other EU regulations and administrative burdens, falls with devastating effect on industrial sectors which are by their nature and unavoidably energy-intensive. The six most vulnerable industries (identified and acknowledged by the European Commission even while it piles on the pain) are: metals, chemicals and chemical products, paper and paper products, mineral extraction, glass products, ceramics and cement.[4]

These are major British industries, constituting a substantial part of British manufacturing – and related services. Tens of thousands of jobs are involved and directly threatened. Many have gone already. And all to exactly what purpose? As in the legendary taxi driver's question to Bertrand Russell in the back of his cab, 'What's it all about, guv?'

A Case for the Defence

Let us for a moment try to defend these two characters and their policies despite the dismal evidence which now confronts us.

Their stance – from 2010 onwards – was broadly inherited from the previous Labour Government and from a remarkable consensus of opinion which had grown up at considerable speed in the first decade of the twenty-first century across all political parties and much of the political spectrum, across much if not all of the scientific community, and across the worlds of

4. The EU's definition of 'vulnerable' industries is set out at www.ec.europa.eu/energy/doc/2030/20140122_swd_prices.pdf.

youthful idealism so successfully tapped by Al Gore and other evangelists internationally. There was, and still remains, a religious fervour in the air about saving the world, ridding it of fossil fuels, and ushering in a new green age. It was in this atmosphere that all political parties in Britain, with a few voices of dissent, supported the remarkable Climate Change Act of 2008, which bound Britain itself uniquely and unilaterally to rigorous carbon reduction targets and timetables, going even further than the EU's own declared goals at the time.

On the European Union side, there was also a strong mood of commitment to fighting climate change as a top priority. Until the Ukrainian storm, as we have seen, this had priority over all fossil fuel concerns. To have gone against it would be to go against the views of most if not all of the governments of the other twenty-seven (later twenty-eight) member states. Whilst there might be doubts, especially amongst the British public, about the infallible worth of policies emanating from Brussels or about dedication to further European integration, to the two energy ministers these would have been anathema. Both came from a party, the Liberal Democrats, and from a wing of that party which regarded even a scintilla of Euroscepticism with horror.

The politics of the 2010 Coalition Government further empowered them to take the course they both did. For one thing, the party to which they belonged, as coalition partners, had to be kept pleased for the Government, led by David Cameron, to survive. For another, and as part of the pleasing process, the entire energy and climate portfolio, and therefore the department, had been placed by agreement, and as a result of intricate bargaining, in the hands of the Liberal Democrats, the true believers in fighting climate change and in keeping Britain uncritically 'at the heart of Europe'. Yet a further consideration was that the Conservative Prime Minister and some of his high command seemed at the outset to be strongly sympathetic to the aims and policies emanating from the Department of Energy and Climate Change, at least until some of the implications of these policies began to emerge.

A benign attitude towards all things green was certainly part of the Cameronian Conservative stance ('the greenest party ever'), accompanied, it seemed, by a less-than-total grasp of the immense complexities of energy policy in all its technical, social, and international dimensions.

It could be argued, when all this background is taken into account, that the two had no choice but to proceed as they did and with the policies they so passionately espoused. But, of course, it could be argued that this is exactly where judgement and understanding of energy complexities should have tempered conviction, but failed to do so. From the very start, Mr. Huhne seemed on an emotional high about decarbonizing the British economy and the thrilling results which would emerge, with hundreds of thousands of new jobs and with a decisive contribution by example to a binding global commitment by all nations to dramatic carbon reductions.

This last aim gained particularly strong support both in learned circles and in the UK Foreign and Commonwealth Office, where an intense commitment to the improbable dream of worldwide legally binding carbon reduction targets remained embedded, having been inherited from the previous Labour Government, and left un-expunged. Somehow, it was fervently believed that Britain, although its emissions reductions might be tiny to the point of invisible by comparison with the great emitters of Asia, Africa, and America, could save the world by its example.

These were shaky foundations on which to build any heavy and penalizing structure of policy. Yet this was the basis on which policies, themselves turning out to be flawed, went ahead.

Power failure: National Grid, blackouts, and UK policy

Advanced industrial economies should not be at risk of power blackouts in any but the most extreme and exceptional circumstances. The ability to anticipate demand and to put in place spare capacity may not be available to the poorest economies of Sub-Saharan Africa but it is certainly available in the UK. The risks of a tightening balance of capacity and demand have been obvious and have been widely discussed for at least the past five years. To have reached the point where National Grid officials are having to issue warnings and tell some consumers that they will have to agree contracts, which allow the supplies they need to be interrupted because of potential shortages of supply, is shameful.

The fault does not lie with National Grid. The problem is a British energy policy which, for the first time in the three decades since privatization, is failing to send clear and consistent messages to investors.

It is important in an age of spin and obfuscation to understand that the risks of a blackout are not caused by a sudden, unexpected surge in demand or by a shortage of any of the basic fuel sources from which electricity is generated.

UK demand for electricity is falling, not rising – the product of gains in efficiency and a flat economy. Supplies of oil, gas, coal, and wind are plentiful, with the world price of all the hydrocarbon fuels falling. To leave the country facing the prospect of 'voluntary' cuts over the winter is therefore quite an achievement.

The problem is that we have barely sufficient capacity to convert the available fuels into electricity. Older power stations, particularly coal-fired ones, have been closed but they have not been replaced by new capacity.

What will happen next? For the immediate future, National Grid – a very competent organization – should be able to keep the lights on. The emergency steps announced to prevent power interruptions and dim-outs, at least up to 2018, represent a rational response to a problem made in Westminster. Supply should be sufficient to meet demand unless something else happens. Unfortunately, things do happen and there are real risks which could still cause major problems.

Since more than 35% of the UK's electricity now comes from gas turbines, the reliability of gas and electricity supplies is intertwined. One regular annual risk is the chance of a winter of really cold weather across Northern Europe which pulls in gas supplies to meet peak demand on the coldest and darkest days, including gas from the UK through the existing gas pipeline interconnector system. The result would be a fierce competition for supplies – particularly cargoes of liquefied gas – and a sharp price spike. This occurred briefly in the winter of 2010.

The Government and National Grid have plans in place including preference contracts with the gas suppliers. The cost could further damage the economic recovery and would require additional support for poorer consumers, but it should be temporary and manageable.

More serious on the electricity side would be further problems in the remaining nuclear power stations, some of which are already beginning to show their age. Safety cannot and should not be compromised. Some stations are already offline and others are being run below their capacity in order to limit the risks involved. More problems would unavoidably create a shortage which would not be covered by the measures announced so far.

The situation presents policymakers with two choices. The first would be to ration supplies. For older voters, this would raise unhappy memories of the three-day week in 1973-4. The second alternative is to reopen some of the recently closed coal-powered stations using the circumstances of an emergency to override the planned reductions in emissions. Four power stations are, as of 2015, paid by National Grid to start up swiftly and come online if things get really tight. They are at Peterhead, Killingholme, Deeside, and Uskmouth. Who foots the bill for this – estimated at about £36 million? Why, of course, the poor consumer!

In summary, the lights in Britain are unlikely to go out for the next three years or so. But one way or another, consumers will end up paying a price for a quite unnecessary failure of public policy.

The best outcome would be if the announcement from National Grid makes people realize that the current approach is not working and that a fundamental change of policy is needed before the situation deteriorates further.

And yet again – Coal Everywhere

In an earlier chapter it has been recounted how, against all odds, coal-burning in Europe as a whole has climbed upwards, with new coal-fired power stations actually being built, new coal mining opening up, and imports by the megatonne being pulled into Europe from cheap overseas sources. This of course is the exact opposite of what is, and continues to be, intended by energy and climate policy strategists – that coal can somehow be phased out as quickly as possible to keep the downward pressure on the further growth of carbon emissions.

As for Europe as a whole, so for the UK as well. The paradox is even stranger when it is realized that most of Britain's coal power stations are under official sentence of death, reinforced by firm pronouncements from the UK Energy Secretary that all 'unabated' coal-burning power stations must be closed. The EU Large Combustion Plant Directive anyway stipulates that all of them should be closed by 2016, and if that was not enough, Britain's unilateral tax on carbon emissions, designed with a price floor of £18 (far in excess of the carbon price set by the faltering trading mechanism across Europe, the Emissions Trading Scheme), was always intended to put a stake through the heart of the coal-burning power sector.

It was widely thought this would happen. Thus Keith Anderson, the chief corporate officer of Scottish Power, warned that the CPF (carbon price floor) 'will make Britain's remaining coal plants largely uneconomic by around the middle of the decade [that is, around 2015-6]'.

Indeed, this may eventually happen under present UK policies. The coal 'surge' could be short lived. Large and quite modern coal-fired stations know they carry the death sentence. To the fury of the Scots, the major Longannet coal station will close shortly. So will Ferrybridge. Others will struggle to convert to biomass, as the colossal station Drax B has already partly done. Whether such conversions make any kind of contribution overall to reduced CO_2 emissions is highly dubious. More carbon could well be generated in the growth, processing, and transport of biomass materials, often imported, than is saved from the conversion – another unintended consequence. It is noticeable that the conversion of further units at the Drax power station to biomass, in the form of imported wood pellets, has been abandoned as 'too risky'.

Yet, despite the drive to eliminate coal-burning from the electricity system today, the proportion of total daily UK electricity supply from coal is still at its highest for many years – around 40%. Previously condemned coal stations are being reopened, new coal mines, being explored, old mines in Durham, being reopened for opencast work. The near-dead have risen up, it seems, in true vampire style from their stake-pinned graves.

How can this be? The answer is not so much that coal is winning and having some sort of renaissance but that other fuel sources are failing and losing. The further switch away from coal to gas, so visible in America, has so far just not happened in Europe or in Britain.

One new source of power generation which was never going to be much help from the start, although receiving lavish subsidies beyond the dreams of any coal-fired station, was wind power. This has been the renewable energy sector which some visionaries actually believed could replace fossil fuel burning but never had the slightest chance because of its fatal intermittency and unreliability.

Studies show that wind conditions around the British Isles are volatile. In very cold, cyclonic weather, no wind blows except in the very north of Scotland. And the same is true in very hot periods. This applies not just to onshore wind pylons but to the huge pylon arrays that have been built and are still being constructed out to sea round Britain. Offshore wind farms can produce electricity at about £155 per thousand kilowatt hour (1 megawatt). The figure may be coming down as techniques improve.

Onshore farms can do better at less than £100, and again the amount may be falling with new technology (and much larger pylons). But these figures are way above the cost of electricity from gas-fired turbines, at about £53, and way, way above coal-fired electricity at about £20, and they, too, have been falling. And of course, all wind installations require gas-fired or coal-fired capacity to be ready to step in when the wind drops but current is still needed.

In the present phase of dramatically reduced world fossil fuel prices, these conventional fossil fuel costs are lower still.

The widening gap in cost between fossil fuel power and wind power has to be filled by someone – in this instance, a mixture of the consumer and the taxpayer. The policy has been to pile the extra costs on consumers, by obliging energy distribution companies to buy a fixed percentage of their supply at the inflated cost required (in line with the EU Renewables Obligation – see previous chapter), backed up by numerous tax breaks. But both these cornucopias have a stopper on them.

The consumer is either the industrialist who simply cannot pay and is finding it necessary to cancel new projects or see them go abroad or lay off workers, or all the above, and the householder who is on the march against sky-high energy costs – and has a vote and a voice.

The taxpayer has another guardian in the shape of the UK Treasury. Officials there were careful from the early days to put a cap on the amount of funds they were prepared to pay out in grants and subsidies for new wind farm building and indeed for renewable energy projects of all kinds. That pot is now all but exhausted.

The generosity of subsidies to wind power in the UK was anyway bound to create a costly excess of planned wind power capacity, and indeed has done so, with pending applications overshooting the pre-announced and EU-designated 2020 target for renewable capacity by 50%. This was yet more evidence that the subsidy level was pitched far too high and the British Treasury was bound, correctly, to call a halt to a thoroughly overheated situation. This it has now done, and legislation has been put through Parliament to cap the soaring wind subsidy cost.

The net effect of these clashes between idealistic enthusiasm for wind power and the hard reality of resources and costs is to make investors in further wind farm construction extremely uneasy, and therefore reluctant. If subsidies are going to be cut, or grants are to be halted, if consumers simply refuse to pay any further charges for super-expensive green electricity, where will the return for their investment come from? Perhaps the good times for the wind farmers are over!

Another deterring threat in the wings came from political quarters, with the former Leader of the Labour Party announcing plans, if elected, to freeze all energy prices, thus making it physically impossible for energy companies to buy high-cost green supply beyond the quantities they are legally obliged to purchase under climate legislation. A more recently chosen Leader of the UK Labour Party goes even further and wants total nationalization of energy utilities, deterring investors even further.

Four factors could yet come to the rescue of the major renewable sectors:

- dramatic technological advances that bring down both capital and operational costs,
- risk-reducing contracts with the state to sell electricity at a guaranteed strike price – the so-called 'contracts for differences' (of which more below),

- a breakthrough in the storage of power, thus nullifying the intermittency snag,
- and, possibly, a real advance in finding cheaper ways of capturing and storing carbon emissions from conventional fossil fuel-burning, as earlier described – although it all looks very remote so far.

The longer-term hopes for lower renewable costs here are not completely unrealistic. The American experience is encouraging. Offshore wind costs remain far above economic levels (over $200 per megawatt hour), but onshore wind costs have declined significantly (down in some parts of the US to $80 per megawatt hour as against $60 for gas-generated electricity). Newer turbines, with increased height and rotor diameter, can capture lighter wind flows than previously needed.

But a basic limitation remains. A national grid system has to have the flexibility to incorporate intermittent current flows of electricity being generated. In the past the generally accepted engineering view was that a stable system could not cope with more than 10% of its supply coming from renewable and intermittent sources. The dream of renewable electricity replacing all fossil fuel sources thus came up against a blunt engineering fact. Digital ingenuity will certainly increase this percentage, so that wind power can make its contribution, and will. Costs will fall further. Technology will find ways of accommodating higher renewable flows without unbalancing the system. But the limitation is always going to be there. Those who believe – and hope – otherwise are placing dreams and hopes well ahead of hard engineering reality.

In the case of solar power, costs have indeed dropped dramatically thanks largely to Chinese innovation and large-scale production – in the process, bankrupting European producers of solar equipment and photovoltaic cells.

This means that the returns from big solar installations begin to look good, even though just ahead lies the inevitable political reaction: if costs are so much lower, then public support, whether from taxpayers or via the obligation to buy solar power, is bound to be cut as well before long.

On the storage front, costs hitherto have been prohibitive and the technology, hesitant. Huge sums have been spent trying to develop low-cost battery storage, without any success. Better and much smaller storage packs may be on the way for households, including Tesla Motors' much trumpeted lithium-ion battery packs. For industrial baseload needs, pump storage projects can help at the margin – schemes which require large reservoirs of water to be pumped up to mountain sites when electric power is flowing in on nice windy days, and then released downhill to drive turbines when the current from wind or other sources is interrupted.

One major pump storage system operates in Britain at Dinorwic in North Wales. Its very existence gives comfort to the power supply system by allowing other generators to operate at slightly nearer their top safe output in the knowledge that a couple of thousand megawatts can always be brought in swiftly (the time estimated is 12 minutes).

But to build this type of facility on a big scale requires big mountains, plentiful water, and lots of money. Norway has plenty of all three, while the Haute-Savoie region of France certainly has the first two.

As for carbon capture and storage, the problem is not just the high costs already mentioned but that the whole procedure involves major associated infrastructure. One plan is to link the piping off and storage of CO_2 in old oil wells with techniques for squeezing out more oil (so-called enhanced oil production). Owners of dying North Sea oil fields, contemplating the enormous cost of decommissioning their platforms, would thus find miraculous rescue by piping in CO_2, postponing the decommissioning day, and enhancing oil recovery.

But these ideas, while they may look good on paper, are still hopes and longer-term aspirations. In the meantime, while coal fills the gap in the very short term, the crying need in the 2020s is for new generating capacity which can only come from gas turbines.

So we come to what might be termed the 2018-19 question. Are these required plants going to be built at speed and in numbers in Britain, and ready in time to provide the cushion which a reliable system needs? The answer is no. Indeed, the reverse may be happening. Some quite new gas-fired stations, already constructed, are being mothballed (as in Germany). This is because with cheap coal, with carbon taxes on gas-fired operations and subsidies still going to wind and solar power, investing in new gas turbines is just not an attractive proposition. In 2014 only two were started where dozens are desperately needed to get capacity up to secure levels.

In 2013 the British Coalition Government planners began to put together a colossally complex set of arrangements which they believed would overcome this reluctance and get the new stations built or at least started. Named Electricity Market Reform, this was going to involve a plethora of guarantees and sweeteners to persuade investors that their money would be safe, and the returns, more or less certain and satisfying. The two central features were the contracts for differences idea, already mentioned, and an entirely new and ingenious new system in the years ahead for offering capacity payments to power generators to keep their plants on standby at all times and to build new ones to do the same thing.

These ideas were in due course embodied in a hefty Bill put before Parliament, and now passed into an Act, which gives the Government powers to pay inducements of two broad kinds to investor or owners of power sources to ensure that power supplies are 'safe' and the cushion against emergencies is sufficient in the years ahead.

But why should energy investors in new and costly gas turbine plants take the risk unless the guarantees are both sufficient and solid beyond any political whim or change of fashion? Belt and braces is what the nervous investor wants, meaning either 'strike price' levels for years ahead or high capacity payments that make it all worthwhile. This is the big fund investor's dream – although it is also the Treasury's, the taxpayer's, and the poor consumer's nightmare.

These new support systems are intended to attract energy investment of all kinds, not just in gas-fired plants, although they are the ones desperately needed in Britain. Ironically, they also incentivize investment in coal-fired power stations and even in diesel power. Thus, as leading British industrialists have pointed out, policy is at the same time penalizing more coal-burning and incentivizing it – a true sign of the inner madness of the interventionist policy mixture.

In short, the aim, in fact, the new necessity in a modern society, is to have a power supply system resilient enough not just to withstand any eventuality like an accident or breakdown, an exceptional cold period, or a terrorist attack. In addition and on top of this, with a growing fraction of electricity supply coming from renewable sources such as wind, and therefore being subject to immense fluctuations depending on the weather, the system must be resilient enough to meet that challenge as well. This is what Britain now lacks.

However, there is one huge low-carbon project which stands to benefit mightily from these support schemes and is neither a wind farm nor a new gas turbine.

This is this large new nuclear power station planned to be constructed by Electricité de France at Hinkley Point in Somerset. There the investors, including both the French and Chinese state companies, have struck gold by managing to get the British Government to agree to charge an amazingly high 'strike price' for years to come – thirty-five years in fact, at an inflation-proofed level starting at £89.50, at least twice the price of electricity from fossil fuels and possibly, in thirty-five years' time, way above the price of any other electricity source including offshore wind farms.

Described by one ex-UK Secretary of State for Energy as 'one of the worst deals ever for the British industrial or domestic consumer',[5] this amounts to a virtual no-lose scheme for the investors with fat returns guaranteed for years to come. In Chapter Thirteen, the saga to date of this amazing impost on the British taxpayer and consumer – in exchange for a nuclear plant design of which no previous model has yet been made to work – will be told more fully. But it shows how incentive schemes of this kind can give outside investors a glorious ride at the expense of taxpayers and consumers.

So how have these elaborate 'rescue' systems worked so far? At the time of writing, one capacity market auction exercise has been conducted. Unsurprisingly, a great many *existing* power plant owners applied, eager to get contracts for the extra payments. The authorities found that a goodly number of suppliers would

5. Lord Howell of Guildford, Hansard, 20 July 2015.

be prepared to join the standby reserve for quite a modest subsidy sum – in fact, £19.40 for every kilowatt of capacity being kept ready, whether operating or not. No less than 49 gigawatts of capacity, about two-thirds of the available total for the UK, has been earmarked in this way, surely more than enough that might be needed if and when supplies to the grid suddenly drop, as they certainly will from all wind sources when, as often, the wind blows too strongly for safe wind pylon operation or too weakly and the blades stand idle.

It has to be understood that all this standby capacity is for some time ahead, not the immediate future. In fact, the aim is to get this kind of insurance against power shortage in place for the winter of 2017-18 (we shall come to what happens in the meantime in a moment).

For that emergency standby capacity three winters ahead, the authorities agreed to pay £1 billion – a sum which will immediately fall on consumers in extra energy costs. For subsequent winters after that, similar costs which may be higher or lower will have to be levied.

Not everyone who applied for these capacity contracts succeeded. Those who failed, including several coal-fired stations as well as some gas-fired plants, may well have to close earlier, making the overall capacity margin problem in the short term even tighter – yet another perversity in the complex energy jungle!

This is where we hit the worrying weakness in the whole scheme. The aim is not just to make existing plant owners a bit happier. It was, and is, to fill the big hole left in the overall power system by rapid coal station closures and get some new, additional generating capacity built quickly. And here the story takes quite a different turn.

The first 2015 auction described resulted in very little new-build so far – in fact only one new combined cycle gas turbine (CCGT), Carlton Power at Trafford, plus a few small ones, 2.7 gigawatts in all.[6] This sounds a lot but was in fact less than had been hoped for. At least 20 gigawatts of new gas plant is needed urgently. Meanwhile, some older gas-fired plants are being mothballed as they simply cannot compete with heavily subsidized wind power. So the safety margin could become still more precarious. An added irony is that a large part of the new capacity payment flow has gone not even into existing gas plants but to the owners of smaller diesel electricity generators, no doubt delighted to have this extra income flow as they struggle in an uncertain market for their supplies and as they burn heavy diesel fuel oil – the most carbon-intensive fuel of all. This is the law of unintended consequences going at full tilt.

The Government were expecting at least 6-8 gigawatts of new gas-burning capacity to be brought forward by the energy companies and their investment backers. But that £19.40 failed to do the trick. They were undercut by a melange of existing operators, of old power stations, and of the small diesel owners.

Did the outcome mean that all these existing energy companies, large and small, were bidding too low and will find in the event, three years hence, that they just cannot afford to keep capacity open after all? Or does it mean that they had suddenly found ways of operating much more cheaply and can manage with lower returns? Or did it just seem like money for old rope – obviously a welcome supplement in uncertain market conditions? And will

6. One more auction took place at the end of 2015, this time awarding contracts for an even more paltry 1.9 gigawatts of new gas-fired capacity, with even more going to diesel generators – the dirtiest form of energy both in terms of CO_2 emissions and toxic nitrogen oxide.

the one new gas-fired plant that to date has taken the 'bait' of capacity payments (a big one called Carlton Power at Trafford) still find that the sums add up for them when construction starts? The answer lies well in the future. It is literally blowing in the wind.

Meanwhile, there is a national electricity supply system to keep intact and resilient for the years leading up to 2018. The technical responsibility for ensuring this falls on National Grid. Under the title of New Balancing Services, they have developed two new tools to help them reduce the risk, both resulting in yet more costs on the consumer but probably just about adequate to prevent power interruptions, brownouts, and blackouts.

One is called the Supplemental Balancing Reserve, which is really a scheme simply for paying older gas-fired stations and one of the last remaining oil-fired power stations to stay open for operation in case there is an emergency. Ways have also been found to keep Britain's ageing fleet of old, 'warhorse' nuclear stations going a bit longer.

The other scheme works at the demand end of the line. This is called Demand Side Balancing Reserve, and is a fancy name for going round to big electricity consumers and getting them to agree, in exchange for discounts, to cut their consumption at times of peak electricity demand, or switch onto private generators.

One way or another, through these sort of patch-and-mend devices, National Grid will, with its immense skill and ingenuity, probably keep the lights on for the next few winters. But the costs will be great, and the danger of collapse, never very far away. It is a measure of the level of policy incompetence and misunderstanding that a nation like the UK should be pushed into such a position.

So far Britain has been lucky. The recession years have slowed the growth of electricity demand. The same circumstances which have checked the growth of carbon emissions (lower economic activity) have also checked the pressure on Britain's electricity-generating capacity. But will this luck hold?

It can now be seen that energy costs are rising not because of market prices (European natural gas prices are at a new low despite the continuing conflict over Ukraine and instability in North and West Africa) but because of these immensely costly schemes needed to keep power flowing and the lights on and because of government support for renewables, which is gradually feeding through to energy bills. These costs will rise much further as more offshore wind and eventually new nuclear capacity – including Hinkley C if it goes ahead – come on stream. Meanwhile, US energy costs continue to fall – to the considerable competitive benefit of US industry and the detriment of companies in the UK and across Europe.

In one narrow sense, all these schemes for keeping power flowing and the lights on in Britain have worked – this is in the fact that the intensity of emissions from the power sector is on the decline and will decline further up to 2020, mainly because of the closure of coal-fired stations. National Grid estimate that it will fall from about 450 grammes of CO_2 per kilowatt hour of electricity to something between 250 and 200 grammes. By 2030 it might be as low as 100 grammes.

That would be splendid if the power sector was the only source of carbon emissions. As we saw in an earlier chapter, this is far from being the case. And it will be splendid if it can be achieved at a manageable cost – again unlikely – and without compromising reliability. But that would require plenty of new gas turbines, plenty of incentives to invest in them and build them, and plenty of cheap gas to supply them.

The Frackers' Dilemma

Shale gas, promised as a holy grail, remains undeveloped in the UK despite ministers' rhetorical support. Falling oil and natural gas prices all round make the prospects for big investment in UK fracking even more remote. As a result, the UK is becoming ever more reliant on imports.

This need not necessarily be bad news. During the second oil shock of the 1980s, the one economy least affected was the Japanese economy, despite having no self-sufficiency at all. Why? Because they drew on a very wide diversity of suppliers of both oil and gas. If one faltered, the others could step in. There was safety in numbers.

The pro-fracking lobby and the Government in the UK have put weight on the benefits that national self-sufficiency in gas might bring for years ahead. The gas is certainly there in the so-called Bowland Basin area, which stretches from Blackpool and Wrexham in the North-West of England to Scarborough in the North-East. A central (and official) estimate is that there is a mind-boggling 1,300 trillion cubic feet of gas deposits in the region. Admittedly, only 10% of this might be recoverable, but it would still be an enormous volume when compared with current total UK gas consumption, which runs at about 3 trillion cubic feet per annum.

But there would be nothing clever about domestically produced gas if it was much more expensive than imports. North Sea gas has been so good for Britain because it has been cheaper as well as more secure. With world gas prices falling and with plentiful gas ready to enter the British market from a variety of sources all around, by both pipe and LNG carrier, the would-be frackers and their backers were always going to face tough competition. Now, with prices being dragged down even further below $40 for oil, the economic case for domestic gas fracking is bound to be even more challenging.

This would be so even if the whole government policy approach to fracking had been well handled. But ill-judged ministerial comments have succeeded in uniting not only the green Left but also the countryside and environmental Right in opposition to fracking developments, making delay and planning caution all the more certain – which means higher costs and lower returns, a certain turn-off for the investor.

Despite extensive sweeteners offered by Government to areas ripe for fracking, the mood of hostility has been hardly affected.

This is because officials began talking about fracking throughout the South-East of England – where ironically it turns out that there is NO gas anyway and very little recoverable oil[7] – under 'old rectories' and in beauty spots across the nation. Most of the commercially recoverable reserves of shale gas and oil lie in the North-West and North-East of England. So the obvious need was to reassure all residents in these regions that fracking would be well away from habitations – ideally at the many sites left derelict after two centuries of industrial revolution, such as disused coal mining areas.

But rather than insisting that fracking permissions would only be given for these suitably remote and derelict areas, which would strongly benefit from new energy investment, suggestions on those lines were brushed aside and derided. Instead, officials tried the line of offering payments to local

7. As confirmed by the British Geological Survey team in June 2013. The BGS spokesman, Mr. Gatliff, confirmed that there was no shale gas in the Weald Region and the geology was 'difficult'. Andrew Aplin, Professor of Unconventional Petroleum at Durham University, thought that 1% of the Weald oil resource might be recoverable – or about 0. 05 billion barrels, less than two months UK consumption.

communities as compensation for disruption, whether during the construction period (endless lorries down narrow lanes) or, in the case of oil operations, the much greater and more prolonged degree of intrusion.

Neither compensation of this kind nor reassuring words remotely match the fears of homeowners faced with the prospect of their dwellings plummeting in value. The result is that hostility from all sides has continued, deterring already nervous investors.

Sweeping announcements from Whitehall that thousands of acres are to be licensed for drilling have raised alarm even further.[8] Proposals that licensed areas should include nature reserves, and even the historic city centre of York, ensure endless opposition and delay, and also amount to sheer political folly. An approach of this kind guarantees that both left-wing anti-fossil fuel protest and more middle- and right-wing concern for Britain's exquisite countryside, as well as fearful homeowners seeing their house values fall, will be cemented together in furious objections and resistance. A clumsier way of accessing gas reserves in the UK could hardly have been conceived. As Spencer Dale, the BP chief economist, points out, if fracking is attempted near houses, it will be entangled in endless cost-inflating delays. He sees gas from fracking in the UK, regrettably, as 'not important'. It could have been. The gas deposits are certainly there at a depth of between 1,000 and 3,000 feet. But he is now probably sadly right. The US now has one million shale drilling operations. Britain in October 2015 had one!

8. In the so-called 'fourteenth round' of licencing landward areas for exploration, via invitations to bidders, along with new tax incentives to make the prospect more appealing.

New government proposals include the provision that no fracking should be allowed within 10 kilometres of specific beauty spots. A far more sensible provision would have been that no fracking should be allowed within 10 kilometres of people's homes. A further guarantee should have been that no fracking should be allowed in or under national parks, let alone in nature reserves. But no such assurance is in place. On the contrary, fracking under national parks may now be permitted, contrary to earlier impressions that national parks would be protected in all aspects.

This insensitive policy approach ensures that fracking in Britain will proceed at a snail's pace, if at all. The more the talk of 'overcoming' local concerns in the name of national interests, the more furious the local outrage and opposition will be from all political directions, and the greater the cost and delay.

The only hope for commercially successful fracking, even if oil prices and gas-equivalent prices rise somewhat, is to do the drilling and build the access roads well away from homes. Two hundred years of industrial revolution has left the Northern England, where the richest shale gas deposits lie, with plenty of unloved, derelict, and desolate sites, abandoned mines and some areas still with a toxic legacy. These would enormously benefit from being both cleaned up and developed, and would be the obvious places to frack, bringing jobs and investment to northern areas while minimizing environmental intrusion. When I suggested this in Parliament as a better course than drilling for oil or gas either near villages and homes or in areas where there was no gas anyway, my words were twisted and distorted in the media wind and met with a gale of outrage and abuse. Had they been heeded, more fracking plans might now be under way. The Northern Powerhouse vision might have received one more boost which it is not going to get.

In Scotland the Holyrood Government has set its face against fracking altogether, as part of an overall Scottish energy policy which guarantees shortages, misery, and high costs in the years ahead. The Welsh Government is Cardiff takes the same line.

So much for the excited talk of nationwide fracking, of gas supplies to last forty years, of a return to the self-sufficiency which once the North Sea provided, and of a new golden age of both cheap gas and national energy security. Under present policies and with present official handling, it will not happen.

******

As for the UK's contribution in the battle to curb climate change, despite all the money devoted to the issue, according to the Government's figures, annual UK greenhouse gas emissions are down only fractionally over five years – a fall which can mostly be attributed to low economic growth – and that is only on the production side. As we have seen when it comes to per capita consumption, feasting on huge volumes of imports from high-carbon countries, the pattern is the complete reverse. Despite carrying a unique (£12) carbon price, coal consumption has risen by 22% over the same period.

Necessary investment is down especially in power-generating capacity because the incentive required to bring in private capital has been driven out of the system. As has been explained, National Grid has had to resort to buying back up otherwise redundant generating capacity in order to ensure that the lights will stay on. We are getting close to a renationalization of the whole system.

None of this could be stated openly in Britain because energy policy, although led by the Liberal Democrats, came within the collective responsibility of the 2010-2015 Coalition. What's done is done, and the only way forward is to find a rational way of unravelling an unhappy mess.

Meanwhile, the price trend for fossil fuels continues downwards, not up, as so many prognosticators loudly asserted. Realists would have recognized from the start that in a hybrid system the role of government is not to be the investor of last resort but to set a framework of policy which encourages private capital to invest in order that the desired objectives can be met in the most efficient way. But realists have not been in control of policy.

Those more recently elected into office must define where they wish to strike the balance. For the last five years since the passage of the Climate Change Act of 2008, the priority in the UK, as in Europe, has been firmly set on the reduction of emissions. Costs have come a poor second. That is why new offshore wind power is being commissioned at up to £150 per megawatt hour, why future capacity is being bought at costs of billions of pounds, and why a new nuclear power station which can never be commercial is being built. It is also why Europe was so unprepared for the possibility of a supply shortage caused by Russian action in Ukraine. The balance now requires some correction. And it is why a two-thirds drop in the oil price and an era of new price volatility, which drags down all other fossil fuel prices, have caught Europe so unprepared.

Rebalancing policy should proceed from realism. The challenge of reducing global emissions cannot be solved in one country or even in one region.

On the cost issue, the answer must lie in competition rather than a price freeze. The fundamental error of the Electricity

Market Reform process is that it is protectionist in nature. Companies are guaranteed revenue, so that the incentives to reduce costs once the deals are signed is minimal. Deals which guarantee index-linked prices – from a starting point way above the current wholesale price for fifteen, twenty or, in the case of new nuclear, for thirty-five years – are indefensible. Each element of the system should be competitive, with performance incentives which allow suppliers and consumers a share of the benefits if the providers succeed in bringing costs and prices down. EMR should be scrapped and a review process, created to come up with the details of a better scheme which can be implemented now the General Election has passed.

UK energy policy has for almost a decade been both the willing victim of misguided and often shifting EU energy and climate policy, and the active and zealous proponent of still more mistakes. The results have been catastrophic. A major change of direction is urgent.

But to achieve that requires further change also in the overarching EU policies which have caused so much havoc. A refocusing of EU energy priorities and aims is equally essential. There are arguments both for more physical integration of energy supply routes across the whole of Europe, and, at the same time, for a substantial devolving of over-accumulated decision-making powers (competences, in the jargon) back to member states. The acclaimed but rarely applied principle of subsidiarity – returning functions to the lowest level consistent with efficiency and effectiveness – needs to be revived and applied. This needs to be a key part of the long overdue reform of the whole EU structure and the renegotiation of the relationship between member states, including Britain, to achieve the successful twenty-first-century EU model for which all, except those most locked in the European past, are now calling and longing.

CHAPTER TEN

Escape From The Labyrinth

The continuing contradictions and stupendous errors of EU energy policy – the possible escape routes – Caspian pipeline possibilities – weak shale prospects – North Sea decline – a policy that might work – switching resources to green energy research and technology – creating a pipeline grid and energy market – Russian gas to Europe will continue – how Britain and Germany should be making a different contribution to meeting climate challenges – prospects for much cheaper green power.

How can Europe, and its most hobbled and damaged states and regions, begin to escape from the truly appalling confusions into which it has fallen or been driven by its unilateral energy and climate strategies? Commentators note how EU energy and climate policy is sliced through with a double set of contradictions. First, there is the obvious ambivalence between concern for energy security and desire to reduce carbon emissions, sharply reflected in the most recent and highly elaborate plans for a new EU Energy Union. Critics from the green camp have been particularly angered by the apparent contradiction between measures to bring more oil and gas safely to Eastern and Central European economies, supposedly threatened by the unreliability of Russian supply, and measures to cut carbon.

Then there is a second ambiguity. The EU wants a transition to low-carbon sources but the EU does not like, and actively opposes, subsidies to national enterprises which distort the Single Market and free competition. So it is simultaneously for supporting green industries and not supporting them. This is the deep fault line running through EU policy.

There should be no disguising (although some will keep trying to disguise) how utterly contrary, self-defeating, and wounding the consequent zigzag policy paths have been so far. Here is one critic, Benny Peiser, the director of the Global Warming Foundation, giving evidence before a US Senate Committee at the end of 2014 about the European scene. He does not mince his words:

> 'The EU's unilateral climate policy is absurd: first, consumers are forced to pay ever increasing subsidies for costly wind and solar energy; secondly, they are asked to subsidize nuclear energy too; then, thirdly, they are forced to pay increasingly uneconomic coal and gas plants to back up power needed by intermittent wind and solar energy; fourthly, consumers are additionally hit by multibillion subsidies that become necessary to upgrade the national grids; fifthly, the cost of power is made even more expensive by adding a unilateral Emissions Trading Scheme. Finally, because Europe has created such a foolish

> scheme that is crippling its heavy industries, consumers are forced to pay even more billions in subsidizing almost the entire manufacturing sector.'[1]

Peiser appears driven by a conviction which is not shared by this book. He believes that the whole climate change issue is nonsense, or at any rate he believes that the argument that it can be defined with precision, with exact dates put on the assignment with catastrophe, and that measures can be introduced to avoid it, is nonsense. It may happen in a century, or a thousand years, or just be part of some great cycle of history, or just be part of changing behaviour in sunspots, or never happen at all. No global warming, he firmly and accurately points out, has anyway occurred for the last eighteen years.

This book's position is short of that. In the words of the late Iain Macleod about Enoch Powell, 'We travel on the same bus as Enoch, but we get off several stops before he does.'

The belief here – and it is only a belief – is that global warming is going to pick up and continue quite fast, that man-made actions have worsened it, and that man-made measures can have some mitigating impact. If they are not doing so now, it is because the measures, not the objectives, are wrong, and because Europe in particular has pursued inefficient policies which are actually making the situation worse – much worse.

How can Europe change its policies and directions, or how can the different national governments which have erred so grievously make good and get back on a more effective and successful transition towards resolving the triple dilemma, the trilemma described in earlier chapters of affordability, security, and decarbonization?

---

1. Benny Peiser, Testimony to the US Senate Committee on Environment & Public Works, 2 December 2014.

Supposing oil prices stay low

In theory, and possibly in the unplanned and zigzag unfolding of events that often march right over the political aims of governments, there are a number of escape routes from the European labyrinth which could gradually bring the overall energy scene back on track and liberate European consumers from the predicaments in which they have been cornered.

The most obvious of these would be a prolonged phase of much lower oil and natural gas prices worldwide. This would have the effect of easing energy prices throughout Europe and making it much less expensive to buy in oil and gas from plentiful world supplies, thus improving security of supply, keeping prices down, boosting economic growth, and financially empowering more communities to choose less expensive alternatives – as well as making more resources available for adaptation for the inevitable weather violence to come.

There is, however, a nasty caveat. It is conceivable that climate policymakers who regard lower fossil fuel prices with horror may persuade governments and the EU to insist on higher levies and taxes on consumers as the underlying price falls. They will argue that prices must somehow be kept high for two reasons. One would be that high energy prices are necessary to force consumers to conserve energy and use it more efficiently. Another would be that unless prices of fossil fuels are kept well up, the subsidy required for green energy sources becomes insupportable, and the extra charges on consumers and on the taxpayer become unsaleable by any government.

One certain low-carbon casualty of a prolonged weak oil price would be the delay in any further significant expansion of nuclear power in Europe. Germany, Austria, Switzerland, and Italy have anyway closed down their nuclear programmes. But plans for new nuclear stations have nonetheless gone hesitantly ahead in Poland, the UK, Finland, France, and the Czech Republic. None of these will be competitive without subsidy, through government loan support and through guaranteed prices for power produced.

As long as fossil fuel prices stayed high, or could be kept high by deliberate policy, and as long as there were general fears about the security of oil and gas supply, this extra cost, both for security and for the low-carbon benefits of nuclear power, seemed worth bearing. But in a world of cheaper and more plentiful energy, this case begins to dissolve.

This is exactly what happened in the UK in the 1980s. At a time of oil price shocks, threats from OPEC to cut supplies, and soaring oil prices, a big expansion of nuclear power seemed a good bet. The French had gone ahead in the 1970s with an amazingly large programme of new nuclear stations. No less than fifty-eight were built, giving the French people 70% of their electricity from this source and a unique degree of energy security. This was done not for low-carbon reasons – no such concern was on the political agenda in those days – but entirely to give France protection against the vagaries and whims of Middle Eastern oil producers.

Ten years later, the UK, which had been the pioneer of civil nuclear power, decided that this was a model to follow. But they were too late. Nine new, large nuclear stations were announced (by this author) and blessed by the Thatcher Government, but oil and gas prices were sinking away fast by the early 1980s, after the rocketing figures of the previous decade. As they fell the economics of new nuclear stations looked less and less attractive. Finally, the programme fell apart and only one station (Sizewell B) was ever built.

The same thing could happen again. This would be a setback for decarbonization but a huge bonus to both domestic and industrial consumers, and therefore to jobs and competitiveness.

The Pipeline Possibilities

Another factor which could improve energy prospects in Europe would be the expansion of pipeline oil and gas facilities from the direction of the Caspian basin. With the South Stream project being abandoned reluctantly by Russia, the opportunity opens up for a new gas pipeline through Turkey if enough gas can be found in the region to fill it and make it economic. This could be a kind of revival of the aborted Nabucco project much favoured by EU members but never started because not enough gas flows through it could be guaranteed.

Pipeline investment and pipeline politics require firm long-term contracts to justify the initial investment, both in the pipeline construction and in security along the route. At favourable prices, these kind of contracts could be attractive to Southern European countries such as Italy and Greece, and further reduce their vulnerability to Gazprom and the Russians.

Pipeline plans for bringing more gas and oil westwards from Central Asia have been reconsidered and reconsidered again as policy spins in the wind of events. The original Nabucco pipeline plan through Turkey was abandoned, and new pipeline routes were planned for supplying South-East and Southern Europe. Tensions between Russia and Western Europe (and NATO), arising both from events in Ukraine and from Mr. Putin's aggressively anti-Western hostility, wrecked Russia's South Stream project under the Black Sea, and European gas importers began to think how they could manage with much less Russian piped gas. Now Russia has proposed new gas routes through Turkey.[2] Europe has cast round for new supply sources elsewhere. Russian leaders now muse about new markets for gas in the East, hoping to make China a bigger customer.

Meanwhile, the EU planners, having rubbished gas expansion for years (which helped undermine the original Nabucco plan), have begun to think that maybe a revived Nabucco pipeline might be viable after all. Such is the on-off-on nature of European energy strategy as it twists and turns in the face of hard reality and unyielding events.

Viking Electricity: Interconnectors to the Rescue

Yet another interesting prospect for Europe lies in a major expansion of both gas and electricity interconnections. This was the possibility of bringing several thousand megawatts of extra electricity by underwater cable interconnectors from neighbouring countries with surplus power (mostly from alternative 'green' sources) which they are ready to export. This major potential source of increased supply from this highly promising source was unaccountably left out of the EMR cornucopia approach at the start.

The idea, embedded in the UK 2013-14 Electricity Market Reform Act, was to reward investors in any kind of new generating capacity (including even coal) with tempting guarantees of long-term payments, whether or not electricity was being produced and sold into the grid system – the so-called capacity payments scheme.

Interconnectors bringing power from elsewhere were not at first included since, went the marvellously official logic, this was

2. Although with currently acid relations between Turkey and Russia, following the shooting down of a Russian warplane, this looks shaky.

electricity produced outside the UK and therefore should not be subsidized by guaranteed returns.

Interconnector systems already exist from France, which has surplus electricity from its remarkable nuclear power system – although only at certain times of the day – and from Ireland which has a surplus of wind power at times – again only available intermittently.

Extra interconnectors are being considered from Belgium, Denmark[3] (overloaded with so much wind power that it almost wants to give it away), second lines, from France and from Ireland again, and long cables, from Norway (900 kilometres), with plenty of spare power, and even from Iceland (1,500 kilometres) which could supply at least 2,000 megawatts from volcanic sources.

The Norwegian and Icelandic supplies would be particularly high-grade because they would be continuous, not intermittent. Norway has enormous capacity to store electricity through hydroschemes, and deliver it whenever required. Iceland's power would come from underground thermal heat sources and would also be permanently available.

Faced with these possibilities, the policymakers have now found ways of extending their sweeteners, so-called capacity payments, even though the electricity being so royally treated does not originate on British soil. The projects are big and quite complex, requiring elaborate onshore switching stations and conversion mechanisms to change the current, which has to travel long distances in DC form, back into AC for delivery to the British grid. And they could take four to five years all to get

3. The Denmark interconnector, actually named 'The Viking Link', would bring 1,400 megawatts of power into the grid system via twin 650-kilometre undersea cables. It would also plug the UK into the Nord pool power market, connecting with Sweden, Germany and Norway.

going, although the Belgian one has already started. They help in the medium term, but not immediately.

These would make a genuine pan-European energy market more of a reality, and would much improve both the security and the distributive efficiency of the whole European system. As earlier pointed out, one of the oddities of present EU energy policy is that while a mass of regulations and increased controls have been drafted into existence to ensure an even energy tableau across the Single Market, in reality, no such market can exist since the means of moving the gas or electricity across the market only exists in very limited areas.

Disjointed energy networks and bottlenecks characterize the system, creating a series of energy 'islands' rather than a comprehensive market.

In Eastern and Central Europe the fragmentation is particularly evident with almost every individual country trying to do its own deal with Russian suppliers. This explains why the former Polish Prime Minister, Donald Tusk, now President of the European Union, has tried – so far in vain – to coordinate European gas purchasers and thus confront Gazprom with a powerful single set of buyers. Prices, he argues, could be set transparently at trading hubs across the continent, thus preventing Gazprom picking off individual customers with secret deals.

A north-south corridor of electrical interconnectors, gas and oil pipelines, transportation routes, and telecommunication links remains one of the most urgent requirements to be fulfilled if the goal of a single European market is to be achieved. It would integrate Central Europe's economies more tightly with Western Europe. Moreover, this economic backbone would enhance European energy security. Its connection to LNG terminals on the Polish and Croatian coasts would allow the whole of Central Europe to tap into the growing global LNG market.

Lack of physical connection is not just an Eastern European problem. Spain has been trying for years to connect up with France and sell its intermittent and highly destabilizing surplus wind-generated electricity to its neighbour. A Pyrenean interconnector is now due to open but it has taken three decades to establish. Attempts to carry current by pylons were repeatedly frustrated and blocked on the grounds that they were eyesores, and in the end enormously expensive conduit tunnels had to be drilled through the mountain range.

For the British in particular, the interconnector possibilities could be a real rescue move – a sort of seventh cavalry arriving just in time to save the British electricity supply system from the ultimate shame and disgrace of interruptions, cuts, and blackouts, arising from monumentally incompetent policies.

All in all, the interconnector potential could play an invaluable part in keeping the British system clear of power cuts and enlarging the safety margin for the moments when things go wrong (e.g. power station breakdowns). The power itself would be both cheap and green, and thereby create a downward pressure on British wholesale electricity prices.

Shale Gas – Another Rescue?

What hopes of greater European energy security can hydraulic fracturing (fracking) and shale gas and oil recovery bring to the scene?

In many ways, its impact has already been immense. Even if the shale revolution never spreads outside the United States (a prospect analysed below), the huge increase in domestic production of gas from shale formations, shale oil, and so-called tight oil (differing from shale oil in that it is mostly found in siltstone and quartz formations although extraction technologies are similar), alters the entire global energy equation.

It is estimated that American oil production will increase by 2020 to just under 12 million barrels per day, placing it well ahead of Russia and almost even with Saudi Arabia. This will come from new fields and so-called 'reserve growth' – that is, extra production from existing fields. American demands on world oil and gas markets are set to fall dramatically.

More broadly, it is estimated that additional production of oil over the next five years from already planned capacity round the world will increase the global total oil output by a minimum of 17 million barrels a day, on top of the existing global output of some 93 million barrels a day. This will be one the speediest leaps in world oil production on record. Even larger increases in natural gas from shale are already in the pipeline.

Looking at this overall tableau of massively expanding gas and oil production now actually unfolding, we are speaking in terms of glut – with major implications for prices, now and in the future (since, as with all commodities, what goes down comes up, often with a vengeance). What this does – and is already doing – to markets, to the fortunes of different economies, and to global geopolitics generally, will be the subject of closer scrutiny in Chapter Eleven.

But will this soaring production pattern be still further added to by fracking output in Europe?

The sober assessment is that the outlook for shale gas and oil development in Europe is not good. A number of factors are working against big shale development either on the continental landmass or in the UK. As we have seen in the UK, badly mishandled policy and opposition from both Left and Right politically have led investors to show every sign of tiring of unending delays and policy muddles, and walking away to other more profitable and certain pastures.

However, the most obvious problem is outside government control. It is the issue of commercial viability. Energy investors of course have to look through short-term oscillations in returns and judge the much longer-term prospect. But how short-term is the present world slide in both oil and gas prices, how prolonged, the apparent glut? Until there are clearer answers, this factor alone will delay fracking drilling and heavy investor commitment everywhere. And this is on top of deep structural and political obstacles. The environmental concerns about fracking weigh heavily on European governments and have led to France prohibiting fracking altogether. These political delays, obviously fanned by hard-line green campaigners who oppose all forms of fossil fuel development on principle, will make progress difficult enough. But rose-tinted rather than frank views about the intrusive impact of multiple rigs, the infrastructure of roads and pipes to get the gas away, the difference between shale gas operations, and the much more intrusive shale oil projects have all helped pile up the opposition and general distrust of official reassurance.

Elsewhere in Europe, the most promising zone has been Poland, mentioned earlier. But here, although fracking began as both popular and strongly government-supported, the tricks of nature and geology have worked against success (as we saw in the previous chapter). Whereas in Northern England shale deposits appear to be both substantial and thick (the Bowland Basin), making recovery much easier, the geological assessments for Poland have been more mixed and early enthusiasm has faded. Meanwhile, German, Austrian, and other European shale prospects have all been approached with great caution and little enthusiasm.

But does any of this really matter? The EU is surrounded by willing and able gas suppliers by pipeline, while further LNG import facilities (to take in plentiful world supplies of LNG) can be built. Some have been already. Norway is ready to pipe very substantial, increased volumes to both the UK and mainland Europe. Algeria is another supply area which, barring political instabilities, can provide big additional piped volumes.

The UK's North Sea Continental Shelf is now past its peak. And strictly speaking it is a global rather than a European resource so far as oil is concerned, as Prime Minister Heath discovered to his surprise during the 1974 oil shock.[4]

But it continues to provide 42% of the UK's daily gas needs and still contains an estimated 25 billion barrels of commercially recoverable oil – a figure which tends to grow as technology increases low-cost access to new fields, and recovery enhancement methods allow older fields to continue producing.

A European Energy Policy that might work

The overall picture that emerges from a review of European energy prospects (and British prospects within the EU envelope) ought to be one of plenty. Even with a phase ahead of Russian unreliability, even with Middle Eastern and North African political turmoil, even with resumed Asian economic growth, the sources of primary energy for Europe, both fossil and renewable, both conventional and unconventional, are

4. According to contemporary accounts, Mr. Heath summoned the leaders of the British oil industry to Chequers at the time of the first Arab oil embargo – 1973 – and asked them to guarantee that they would keep Britain supplied from the North Sea on a priority basis. He was surprised to be told that no such guarantees could be given. The North Sea, he was reminded, was an international province. Supplies would be sold on a commercial basis to customers anywhere. There could be no territorial restriction.

basically plentiful, accessible, affordable, and consistent with a gradual trend towards lower-carbon emissions.

Yet that is not the picture on the screen. There we see something quite different: shortages, eye-watering price increases, fears of interruption, loss of jobs. This is the staggering contrast between what ought to be and what is.

The repeated claim of climate change experts and their committees has been that the climate danger can be met through reducing emissions 'which we can do at reasonable cost'.[5]

Yet the cost has ceased to be anything like reasonable, and the results have not been achieved.

Can a revision of EU energy policy, or of the policies of component member states, bring the two paths closer together? Is there a European energy policy that dovetails with global developments constructively?

An ambitious attempt to resolve some of these contradictions and outline a new way forward in European energy has been put forward by the European Commission. It is entitled the European Energy Union. Offering something for everybody, it tries over many pages of print to reconcile Europe's green ambitions with the crying shorter-term need for better fossil fuel links (i.e. gas pipelines) and electricity interconnections between the twenty-eight member states. Progress on interconnectors would indeed help create something nearer to a genuine market for energy supplies than at present exists, and reduce the delusional element in the Commission's energy planning referred to earlier.

But the gap between plan and achievement is enormous. Member states governments remain deeply reluctant to surrender

5. Lord Deben, Chair of the UK Committee on Climate Change.

their national energy policies to a common European standard or to make the harmonizing changes which would allow gas and electricity to flow freely across borders. There is also the unanswered question of who would finance the criss-cross pattern of pipes and transmission lines required. More awkward still is the dichotomy running through the whole Energy Union scheme between the need to ensure reliable fossil fuel energy supplies to Eastern and Central European states and the high hopes (now not so high) for a grand European transition to low-carbon energy – a divide which green groups have been quick to point out and which led one involved Member of the European Parliament to label the whole scheme as a 'ragbag advocated by schizophrenics'.

But apart from these inevitable slings and arrows between the two camps – the hydrocarbons versus the low carbons – there is an even bigger and more damaging omission – the matter of cost. Cost has been the orphan of European energy policy. Yet a far greater emphasis on getting the costs of energy down could actually be the best contribution at European level to resolving the great energy trilemma, of bringing together the triple objectives of affordability, reliability, and reduced carbon emissions.

Had this been realized earlier, then the central focus all along of EU energy policy would have been on the technology, the science, and the innovation which would have made every link in the energy supply chain less expensive and which would have combined green and renewable power sources, in all their obvious plenty, with cheaper, not much more expensive, energy. Clean coal in all its inevitable global ubiquity would be just as much part of that chain as other sources.

Yet that is not at all the way that the EU authorities have viewed these matters. On the contrary, recent attempts at formulating a common energy policy make no mention at all of cost, innovation, or the impact on European competitiveness and therefore jobs.

In June 2014 the European Commission issued a working document entitled *In-depth study of European Energy Security*. This is a good example of EU energy thinking. The agenda priorities here are the security of supply and the reduction of carbon emissions. Cost does not come into it. The argument instead is that the EU, ignoring costs and competitiveness (and strains on household budgets), should maximize its own energy production.

As the perceptive analyst and columnist Nick Butler has pointed out, there are strong echoes here of the Common Agricultural Policy – the central tenet that domestic production of all kinds, disregarding competitiveness, can and must be subsidized in the name of protection and security. The punishing costs of all this, as well as the obvious inefficiencies of blocking out cheaper imports, are disregarded.

As he reminds us, the original Common Agricultural Policy was conceived against a background perception of food shortages. Similarly, current EU energy thinking, jolted by Russian behaviour and by the fact that Russia supplies 27% of the EU's daily gas, is that there is a shortage of energy supplies.

But, of course, there is no such shortage. Fossil fuels production is rising worldwide, while non-fossil sources round the world, if not in Europe, are growing cheaper. Even nuclear power may find ways of curbing its enormous capital costs and becoming the 'cheap power for all' that its pioneer founders dreamt about (see Chapter Thirteen).

Revised and effective European policy in these fields would concentrate on three aspects – on all-out support for technologies to increase energy efficiency at every stage from production to consumption, on the rapid completion of a pan-European interconnecting infrastructure to allow energy (gas and electricity) to flow freely, and on maximum devolution of energy decisions and policies to member states.

The first would dramatically bring down all energy costs from the producer end to the consumer end. The second would ensure that the benefits of plentiful cheap energy are available in all corners of the market. The third would give much greater freedom to member states to pursue their trilemma policies and determine their own ways at their own pace of cutting carbon emissions and transforming their power industries. The long neglected principle of subsidiarity[6] should long ago have replaced the EU's centralizing tendencies.

Thus, a truly commonsensical European approach would home in on new gas pipeline supplies into Europe from the cheapest sources, new LNG terminals where costs are justified, and a greatly enhanced network of both pipelines and hubs within Europe to create a genuine market, to bring supplies rapidly to Eastern and Central European states, who are over-reliant on Russian gas, and to ensure that there is a real energy market to operate (and regulate).

To switch policy in this new direction will require changes at both national and EU levels.

In Germany the full force of government support will need to be turned from ballooning subsidies for renewables operations, and the crippling surcharges on consumers needed to finance them, towards new technologies and the science underpinning them. Bringing costs down, both for new alternative power sources and for conventional fuels, needs to be the central focus

6. The organizing principle of the EU that matters should always be handled by the lowest level competent authority – a principle extolled by EU planners and policymakers but largely ignored.

rather than the neglected side issue. The German nuclear power decision will have to be reversed, with electricity from already built nuclear plants, where capital has already been sunk, flowing back into the system both cheaply and with minimum carbon emissions.

Austria and Switzerland should follow the same example as Germany's 'Energiewende' swings back towards constructive reality.

In Britain the same kind of changes are urgently required. Energy prices are much too high, investment in gas turbines to replace ancient power stations, too low. Rising subsidies to high-cost wind power need to be more severely curbed as subsidies to solar power developments and onshore wind farms already rightly have been. Any further 'deals' for new nuclear power stations need to wait until new-build costs come down from the inflated levels involved in the Hinkley Point deal described in Chapter Nine.

At the EU level, a new mixture of both lightness and firmness is urgently needed. The lightness should come from minimizing the EU's scope to dictate in the matter of the energy strategies and energy mix of member states. The firmness has to come from pan-European policies to get the necessary interconnectors completed, running from Portugal to Latvia, and to encourage the right diversity of economic supplies, both piped and frozen, from the surrounding suppliers – in the Caspian region, in North Africa and the North Sea and Norway. It is conceivable that the Eastern Mediterranean may also be able to supply commercially priced gas into Southern Europe. The deposits are certainly there, to the south of Cyprus, off the coast of Lebanon and Israel, and to the north of Egypt. But getting them out at a competitive cost still looks a long way off – except in the case of Israel.

The biggest supplier of gas to Europe may continue to be Russia. The difference will be that its pricing policy will be up against much more competition, and the customers, while happy to take piped gas at good prices, will no longer be so riskily reliant. It is reliance which is the key, and the premium that customers will pay for not having to rely too much on one supplier which is the question.

The Caspian Sea region can certainly be brought into the European supply pattern more effectively than in the past. Earlier EU attempts to develop a 'southern corridor' to bring gas from Azerbaijan ended in failure, with the original, much vaunted Nabucco pipeline project having to be abandoned. A new pipeline chessboard has unfolded.

This involves bringing gas from both Azerbaijan and Turkmenistan into Europe via the Trans-Adriatic Pipeline (TAP), partly owned by BP and running right into Italy. Whilst Nabucco One (the dreamchild of the EU leadership who discussed it while all attending Verdi's opera in Vienna) failed because there was not enough gas available to fill it, this time a smaller and more carefully planned pipeline project could be made to work and be up and running by 2019. That, at any rate, is the hope of Maroš Šefcovic, the most recent European Commissioner overseeing the energy agenda.

Because Russia looks set to continue as Europe's biggest supplier of natural gas, she will now also be on the Turkish scene and will be ready to push gas through into Southern Europe. The Russians, having given up the plan to build a trunk gas pipeline directly from Novarissk to Varga in Bulgaria, directly under the Black Sea, now plan to take their gas again under the Black Sea but this time to Turkey on the Asian side of the Bosphorus. From there it will seek markets, as before, in Europe. However truculent Russia may currently be in

Ukraine, and however much EU planners may long for freedom from reliance on Gazprom, the Russian role in Europe's daily gas supplies will remain central and significant. From that there is no escape. A difference from the past is that Russia may eventually, and despite the recent stand-off, find in Turkey an increasingly cooperative friend, both on energy matters and other issues. The Turkey of yesterday with its Westward leanings and its ambitions to join the European Union has gone. A new and less Western-committed Turkey is emerging, which sees itself as both a key energy hub and, more broadly, a major power in the Middle Eastern region.

******

In the round, European unilateral energy and climate policy to date has turned out to be a stupendous error, and a costly and hurtful error at that. Had Barbara Tuchman lived on into the present age, this could well have been one of the milestones in the onward march of establishment folly which she sketches out.[7] At root, Europe's well-intentioned policymakers have been fed with bad guidance leading them to believe that greenhouse gases can be controlled, and global warming, curbed relatively cheaply, straightforwardly, and quite swiftly – and centrally. It turns out that the costs of transforming energy systems to more sustainable patterns are proving far higher than predicted by experts, and the gains, far more ambiguous and indefinite. 'Transformation',

7. Barbara Tuchman's famous book, *The March of Folly: From Troy to Vietnam* (Michael Joseph, 1984), depicted a number of the moments in history when governments, peoples, and opinion became unanimously united in a project or adventure which turned out disastrously. Everyone believed it was reasonable and right. Yet everything turned out to be nonsensical and wrong.

with its undertones of complete fossil fuel abolition, may be entirely the wrong concept. A more balanced approach and a shared future for both fossil fuels and green energy could be by far the best way forward.

A major culprit here has been the widely acclaimed but deeply misleading Stern Review, published in October 2006.[8] This argued that serious measures to avert global warming could be put in place relatively inexpensively if early action were taken. But the costs which consumers have been asked to bear, even before much has been achieved, have already proved much higher, and the persuasive powers of the message, much weaker than the optimists or the supposedly expert analysts hoped for.

Current estimates, for instance from the International Energy Agency, of the infrastructure costs requirements to bring about a low-carbon global economy run off the page with zeros. With present policies, the attempt to meet expected demand, to replace declining production, and to install new low-carbon capacity and supporting infrastructure, needed to meet official emissions targets. The IEA lists a figure of $48 trillion over the next twenty years, between 2015 and 2035. How such sums are to be raised and from whom – about three times the total product of the entire planet – defies imagination. The attempt would impoverish and destroy on a cataclysmic scale. It might well not produce the emissions cutting results sought. It would also, by the way, require an embryo world government to see it through.

As argued in a book I co-authored with Dr. Carole Nakhle in 2007,[9] the appeal to consumers in all nations to pay painful extra

8. Sir Nicholas (now Lord) Stern, *The Stern Review on the Economics of Climate Change*, October 2006.

9. David Howell and Carole Nakhle, *Out of the Energy Labyrinth: Uniting Energy and the Environment to Avert Catastrophe* (I. B. Tauris, 2007).

costs to save the planet at some distant future date would not be sufficient. The message was wrong. A far more compelling theme would have been that it was future energy security that necessitated a shift to low-carbon power and that green technologies, as well as nuclear power facilities, could in due course produce cheaper instead of painfully more expensive energy supplies, as well as safer and more reliable power sources. The Al Gore appeal – that saving the planet requires sacrifice now, and that the science of imminent doom is settled beyond dispute – was never going to carry enough people with it – certainly not when they were told the true cost rather than the misleadingly small figures. Vague, weak, and implausible messages were always going to produce bad results, and this has proved to be the case. Intense hostility has now built up against the green cause.

A Coalition Government in Britain which wanted to be 'the greenest government ever' found itself on the defensive and on the retreat over its energy pricing policies. Two further problems are still weakening the green case.

First, it is evident to the most casual observer that Europe's 'example' to the world's biggest carbon emitters – notably China, the US, and Russia – is having no effect whatsoever. Despite strong concern within the Chinese hierarchy about pollution, especially sulphur pollution from coal-burning, CO_2 emissions continue to soar, adding more each year than Britain's *total* annual emissions. Ironically, the US is achieving big reductions in the pace of carbon emissions growth (but not of course in actual total emissions which remain the world's largest after China). This comes about not through taxes and high-cost energy, not through following the European example, and not through any kind of commitment to the protocols originally agreed at Kyoto in 2003.

Instead, the drivers have been technology, innovation, and cheap power. China's green achievements, such as they are, and its commitment to renewable energy sources, mainly wind power, are also being driven almost entirely by the wish for greater energy security and the desire to be independent of outside supplies. If any example is being followed in Beijing, it is more the American than the European one. If Europe's example is having no influence, why, it is asked, do the extra burdens have to be born? What is the purpose? Where are the results? Why does so much diplomatic effort need to be expended in seeking the holy grail of a global commitment to legally binding carbon emissions targets when no such goal is remotely attainable or needed?

As each international gathering takes place to monitor world progress in the battle against destructive climate change, these are the questions that get asked but receive no answers. The battle is being lost through political incompetence, poor messaging, and seductive but implausibly low-cost estimates – and everyone knows it.

People will make sacrifices and accept extra burdens on their budgets and family lives only if they are persuaded that the benefits are real. But in this respect the scientific community and its supporting lobbies have let the green cause down badly. The author of this book is no climate 'denier'. He believes that terrible climate violence could be coming, that, amongst the many causes of this, man-made greenhouse gases could be one, and that preparations and steps are possible to mitigate, adjust to, and just possibly delay the threat.

But the spurious precision given to the prospects, and the dictatorial tone in which forecasts are handed down, have all added to the backlash. Green extremism and the over-hasty rush to decarbonize have together all but destroyed the green case.

Underhand methods by organizations such as Greenpeace, of which this author has personal experience, and intolerant zealotry in the face of scientific hesitations, have all further compounded the doubts about sustainable energy measures and whether the costs are worthwhile.

Britain should be campaigning vigorously to change the direction of the European Commission's ineffective and catastrophic policies.

Successive EU energy and climate packages have cost the whole continent dearly. The UK Government's original, ill-judged impact assessment of EU policy in 2007 which included the rush to renewable energy sources almost regardless of cost, estimated with absurd optimism that it would involve a recurring annual cost of £3.4 billion but with benefits of £20.4 billion. Wishes became forecasts. Almost none of the benefits materialized. They depended on a global deal on climate change being struck with Europe at the fore, and on soaring oil and gas prices which less expensive and cleaner green energy would replace.

As was clear almost from the start, none of this was going to work out. Now, almost a decade later, and at huge cost to Europe's households and businesses, it has begun to sink in that the current policy is inadequate and damaging.

The clear lesson is that long-term, prescriptive, target-dominated, inflexible, top-down energy policies are not the way to go. Flexibility, diversity of approach between states, and constant review must replace the rigidity of European centralization.

The Realities No one wants to Face

Looking out across the whole European energy scene, certain realities jut out into a swirl of illusion and confusion, like jagged rocks into a heaving sea.

When all is said and argued and debated and done, in the battle to curb greenhouse gases and combat climate change, the real truth which few like to admit or confess is that the contribution made by European nations makes very little difference. The struggle to limit carbon emissions passing into the atmosphere will be won (or lost, as it is being at present) in India, in China, and in Russia – and also in America, where annual emissions are falling but are still by far the highest per head in the world.

The harsh and self-harming measures being taken by Western European powers to impose very high energy prices on homes and industries, and to switch large swathes of income from hard-pressed families and struggling workplaces to big investing corporations and wealthy landowners, even if they are effective – which they are not – add nothing. The numbers are utterly insignificant.

The contribution which countries such as Britain or Germany should be making to halt climate change is quite different from the one they are making now. The billions being poured into subsidies for uneconomic renewable energy – and, in particular, into vastly expensive wind farms – should be concentrated on developing the new technologies to get the costs of greener forms of power production down, to burn fossil fuels more cleanly, and to deliver the massive flows of cheap electricity which developing nations are absolutely determined to have.

The argument that European efforts, while marginal in size, will somehow influence the really big emitters by example has proved a broken reed, despite the string of global conferences and wordy undertakings all round. These huge societies have their own agendas and their own priorities. They will always go for the cheapest energy source – which is at present mainly coal.

The failure to deliver the methods by which coal can be burned far more efficiently is a certain guarantee that emissions targets will be missed and that climate dangers will worsen. The steps taken instead in the West to penalize all fossil fuel use, to deter investment in all energy companies, and to prop up uneconomic sources, in effect, betray the climate cause. They are steps backwards, not forwards, leading, as has been shown, to more coal-burning, not less, higher and not lower-carbon emissions, more hardship, lower growth, and more barriers to green progress. They are a hideous diversion from the real task.

Dreams and Hopes

The world of power from oil and gas will only eventually be effectively challenged by green and renewable energy sources, not because they are favoured by government policies but because they become *cheaper and more reliable, and because they are what the consumer wants.*

When new technologies – which may not be far away – begin to make the generation of power from green sources, its transmission, its distribution, and its daily use straightforwardly cheaper and easier for consumers than power from oil and gas, the market will do what countless government measures and schemes will never do. It will turn consumers, as it always does, to cheaper and better methods. If these include nuclear power as well, or abated coal-burning, that is what will happen. If the nuclear engineers cannot get their costs down, then they will not find investors or pliant governments and taxpayers, and other lower-carbon energy sources will lead.

Home-made power for households will become the practical and most economic method. Localized microgeneration will become the more reliable source. Transport will look to green-sourced electricity as the cheapest fuel. Industrial plants will go for the least costly sources, whether local or imported. The days of large-scale, centralized systems of power generation and transmission, as well as of oil processing and product distribution, will be defeated by the simple fact of cost, with behind it the anything-but-simple new powers of the digital revolution in all its aspects.

After all, digital power has driven down the cost of information over three decades by about one hundred times. The same power might well in due course – perhaps not so dramatically – drive down the cost of delivering and using power. The energy policies of yesterday will be redundant, and the effects not just on the geopolitics of energy but on the whole pattern of global power and influence will be colossal.

This will be the fourth and maybe the ultimate energy revolutionary force of our times and our planet, following on from the oil shocks of the last century, the rise of climate concerns, and latterly the shale oil and gas phenomenon. Indeed, the earth-shaking process, the fourth transformation, may already be under way.

To this we will turn in the final chapter – but there are some crucial steps along the route which come first. And are we on the right route at all?

******

CHAPTER ELEVEN

The Price

How the crude oil price decides everything – the predicted upward spiral in price fails to appear – oil price volatility throughout history – price forecasts usually wrong – price collapse and the Middle Eastern producers – the link with the jihadis – world oil supply abundance – OPEC and the lack of Saudi policy choice – false hopes for non-OPEC 'responsibility' – estimated break-even points for oil-producing countries – the geopolitical consequences of weak oil prices – are lower oil prices better for both producers and consumers?

We had to wear masks. The air was thick with black particles and choking fumes. From the nearby well heads we could hear the roaring noise of burning, spouting, flaming oil.

Further away, across the black, smoke-filled landscape, we could see the spurts of flame dotted along the horizon – something straight out of a mediaeval or renaissance painting of a burning hell.

This was the giant Burgan oil field in Kuwait as the retreating Iraqi army had left it a few days before. Apparently, their saboteurs had only appeared at the last minute as the troops pulled out. It was an act of vicious spite and vandalism – a taste of the descent into all the Middle Eastern barbarity and destruction which lay ahead. We were watching our own energy future go up in flames because this was oil and gas foregone, oil supply shut off, less oil and higher and higher prices. It was bound to be.

That was 1991, days after the end of Saddam Hussein's impulsive invasion and attempt to annex Kuwait. Yet on the oil front, this was not quite how things turned out. The engineers in a resurrected Kuwait moved in and with staggering speed got the oil flowing again. Saddam was put firmly back in his box (for a decade or so) and oil prices, instead of soaring further, settled below their high points – at around $80 a barrel.

The one correct prediction was that there would be more violence and more blood – much more. Oil wells continue to be blown up, production, halted, pipelines, sabotaged, refineries, attacked and staff, slaughtered in Syria, in Iraq, in Libya, in Algeria – the list goes on.

Yet the forecast spiral in world crude oil prices, the shortages, the interruptions which growing Middle Eastern chaos and violence was sure to bring, the blood-curdling talk of oil at $200 or $300 a barrel – none of this has occurred. On the contrary, oil prices have first stabilized and then sagged and now slumped. The 'certainty' of sky-high oil prices, repeated and predicted by numerous analysts and experts, has been overtaken by the reality of sagging prices and surplus oil. Onshore inventories are full; offshore moored tankers are filling up. The old, old pattern of 'too much oil' has come back to haunt the oil industry as it did in the distant past.

The crude oil price is not usually the most widely followed index. Yet in terms of impact – on daily lives, on world events, on the play of power, and on the geopolitics of a turbulent world – it is movements and trends in the daily crude oil price indices which could be the most decisive of all. Despite the vague (and inaccurate) perception round the world that oil's age of supremacy is over, that oil production has peaked, and that oil demand is in long-term decline, it is the quoted price for 'black gold' – crude oil – in the great world oil markets (usually two 'marker' figures either side of the Atlantic – Brent crude and West Texas Intermediate) that continues to decide almost everything else.

The reason this is uniquely so is because movements in the oil price, especially big ones, have shattering effects on both the producer and consumer sides of the equation. That is to say – as touched on in Chapter Two and in much more detail later in this chapter – that when the revenues flowing to the governments of oil-producing states round the world shrink or rise, the impact directly affects their power, influence, and weight internationally, and ALSO their internal social and political structures and general stability. Governments and ruling cliques feel safer and maybe more aggressive when the oil and gas revenues flow plentifully, and weaker and more dangerous when revenues start to shrink.

And when the price of crude changes significantly, the consuming nations find their fortunes directly affected, up or down. Consumers, both industrial and domestic, react in two ways. First, as political actors, they object to higher prices, demand protection, call for subsidies, and so on. Business delegations press the politicians; household consumers change their voting allegiance or take to the streets. When they fall, it is the opposite. Economies are given 'a shot in the arm'. People feel quietly more confident and under less spending pressure.

Second, consumers switch fuels when oil prices rise – and switch back when they fall, although, as we shall see, this pattern of behaviour may be changing as weary and wary consumers move away in advanced societies from heavy energy consumption to a pattern of permanently lower demand. This has a direct knock-on effect on demand for other fuels and for energy-saving devices and technologies. Thus, a big oil price hike creates an immediate increase in gas demand, 'a dash for gas', which in turn flows back into the geopolitical loop, increases pressure for finding other resources (such as shale gas from fracking), makes other energy sources, such as nuclear power and renewables, look more attractive, and generally sends waves of change through the whole energy mix. A big oil price drop has the reverse effect, dragging down gas prices and flooding the market, while deterring investors from committing their money to new oil and gas projects.

Oil is a commodity, and like other commodities, its price is determined at any one time by supply and demand. Like other commodities, it is subject to big up and down cycles, with low prices deterring new investment and new production, thus creating a shortage of supply and hence higher prices and restored investment and so on *ad infinitum.* Equilibrium is not a natural state any more than in financial markets.

The difference with all other commodities is that both the supply and demand sides of the equation are shaped by highly political factors, decisions, and events. On either side, the key questions are what and who determines supply levels and what and who determines levels of demand.

Over the last forty years, crude oil prices have moved up and down over a staggeringly big range. They were $40 in April 1982 (historical prices) but $10 by May. In 1986 they started the year at $30 but were down to $12 by April. In May 1999 they were down to $17; in October 2001 they were $29. In January 2002 they were down to $20. In June 2006 they dropped from $140 to $40 in a month. In February 2009 they were $39, but back to $120 in 2010. In early 2016, they were hovering around $30.

On each occasion, the question has been how long they would stay down and when they would come up again. Sometimes the process waits for normal boom-bust-boom market cycles to operate (lower price equals less investment, which equals less production, which equals shortage, which equals higher price). Sometimes an outside so-called 'high-impact' event, such as the Iraqi invasion of Kuwait, or a terrorist attack on a key installation, crumples all trends and predictions, and sends the price racing along another path.

A final example of the problems of accurate oil price predictions, even in the hands of the most experienced oil experts and practitioners, comes from the author's own experience. In the summer of 1980 I was shown into a special research room at Shell's then Head Office at Shellmex House on the Strand.[1] On the wall a large graph showed Shell's view about the evolution of oil prices over the next ten years – to 1990. There were three lines. The lower one showed oil reaching $60 a barrel by 1990, the middle line reached the $90 dollar figure. The top line touched $120 when the ten years were up.

There had been deep debate within Shell about which forecast made the most sense. In the end, perhaps predictably, the consensus had closed round the middle line. $90, rather than the lower $60 or the much higher $120, these expert and thoughtful analysts believed, was the most likely figure.

1. Now the HQ of Penguin Books.

On that same day ten years later in 1990, the price for Brent crude stood at $18!

Hey-ho, it's a tricky business. But let us at least have a stab at it. Cataclysmic events aside, like, for example, large-scale upheaval in Saudi Arabia, the question now is whether the world is in a normal 'down price' period which will correct itself in a few months, or whether there are some entirely new factors on the scene which will severely prolong the time of weak and falling prices, and severely prolong the impact on the world economy maybe for a decade ahead. Many experts and analysts predict the former. The price, they argue, will bounce back soon, say to $75. This book's belief is cautiously the latter – and for the reasons, facts, and suppositions which now follow.

The Old Order

All energy issues, it used to be said and was widely believed, turned on the Middle East. This was the region with most of the world's recoverable oil reserves (66%) and therefore the region whose stability had to be of profound concern to all advanced industrial nations, the Western powers in particular. Hence the intense and intimate involvement of 'the Powers' in Middle Eastern affairs throughout the twentieth century as their dependence on imported oil increased.

Hence also the realization amongst Middle Eastern producers that they had a monopoly potential in their hands, a realization which about two-thirds of the way through the twentieth century turned to hard policy with the formation of the Organization of Petroleum Exporting Countries – OPEC.

For the preceding half century, this was the development the industrial nations had feared. Ever since the oil age had begun and as factors like the switching of the colossal British

Navy from coal to oil (in 1914) began to take centre stage, the foreign policy priority of Western powers had been the control of Middle Eastern politics and therefore of the oil flow. Britain and France jousted first with the Ottomans – and then with each other, for regional supremacy, while T. E. Lawrence with his pan-Arab vision complicated the scene further (first with his aim of defeating the Turks – which proved successful – and then with his attempt to squeeze out the French – which proved unsuccessful).

The Ottoman withdrawal from the whole Middle Eastern and North African region left chaos, and the chaos of course has continued. As Arab nationalism took over, France and Britain were pushed out and extruded from the Middle East – Britain, after bloody encounters with Israeli terrorist tactics, ejection from Iraq, defeat at Suez, and a general decline of influence. American influence waxed and waned, although involvement through the Israeli connection still remains impossible to untangle.

Now switch the template. Supposing the Middle Eastern region ceases to be so vital for world energy supplies, at least to the West. Suppose the global demand for oil sags, crude prices retreat, the US cuts its imports drastically thanks to its new-found shale oil and gas plenty, other non-OPEC oil and gas resources are brought into a commercially recoverable phase thanks to new extraction technologies.

The outcome would be two-edged. The great oil-importing countries would be able to congratulate themselves that they no longer need to wade so deeply into the Middle Eastern quagmire. Bases could be closed, force commitments, run down, other more reliable energy sources, encouraged.

But what about the consequences in the Middle East and for OPEC, the hitherto all-powerful producer cartel? The oil weapon would fall from its hand. The old power to turn the oil taps on and off and thereby affect the world crude price directly, would slip away.

Why? Because as explained back in Chapter Three, any cut in production by Saudi Arabia or Kuwait or the United Arab Emirates – the three OPEC producers with the biggest oil production and reserves – merely provides space in the market for non-OPEC oil producers to sell more. Nothing happens to the price any more if Saudi Arabia cuts back. All that would happen is that they would lose a further share of the market. A thousand small shale operators in the USA would bounce back into business. Revenues would fall, weakening governments and their capacity to seal off and suppress unrest with money. The wealth of Croesus would shrink, and every kind of tribal bitterness would break out. Deep religious divisions would split societies, smash boundaries, release countless tribal rivalries and grievances, and make nonsense of the geographical order imposed either by the old Ottoman Empire or by its Western successor states.

So even while the joys of reduced oil import dependence would be absorbed into advanced industrial economies, a new threat of equal or greater toxicity would spread out from the region. Sunni and Shia divisions would burst out into widespread civil war throughout the Middle East, with extreme jihadist groups pursuing unlimited violence and extending the reach of terrorism into the soft centre of Western societies. What might have been localized warfare in a previous age would take on a global dimension thanks to the immense amplifying power of information technology, weapons technology, and the possibility of nuclear weapons falling into terrorist hands.

A weakening oil price would also change investor perceptions, cutting the flow of investment funds and luxury purchases from oil-rich countries into the faltering Western economies, but also cutting the readiness of investors to put their money into expensive Arctic exploration and other new frontier energy projects. It would also send tremors through other oil-rich economies such as Russia, where the political elite floats on a sea of vast oil and gas revenues. It would become more important to Russia to keep customers for its gas and to make up lost revenues, thus altering its relations with both Western European and Eastern Asiatic customers.

This would in turn affect, although slowly, the Russian stance over Ukrainian issues, maybe modifying Vladimir Putin's aggressive impulses in the area. Napoleon is said to have been defeated in Russia by 'General Winter'. In the end, it could be 'General Oil Price' who will defeat Vladimir Putin, although not quickly and not without bitter internal struggles and external repercussions.

So what are the price prospects? Price, it needs to be repeated at the risk of banality, is set by the interplay of supply and demand. On both sides of the equation, very big changes are at work, shifting the intersection between the curves on the graph. It takes someone brave or foolhardy to be definite about future price trends. Politics can invalidate economics overnight, as our story from Shell in 1980 illustrated. One violent 'high-impact' incident can change the whole prospect. But here is one set of views, supported by both facts and past events, which this book contends; barring earthquakes, either natural or political, it fits most closely with both economic and political realities in the short and medium term.

The Growing Oil Surplus

First, on the supply side and contrary to many peoples' expectations, we are entering a period of almost unprecedented expansion in oil supply capacity.

Adding together all the absolutely firm country-by-country prospects for additional production capacity (which we examine below), and making allowances for existing well depletion rates, the most sober estimate is that by 2020 the world could be producing (note the 'could') a net additional 17.6 million barrels of oil a day, on top of existing production of approximately 93 million barrels a day.

Persisting low prices could shave these figures marginally, but for the most part they involve committed projects in the pipeline, or in areas where capital costs have already been sunk and it is only operating costs (that can be very low indeed, even for America's frackers) which are affected.

Even this is not the whole basic supply picture because technology is racing ahead to deliver new methods of extracting oil and natural gas liquids from existing reservoirs at breakneck speed. This means that depletion will be offset by so-called reserve growth (i.e. more oil production from increased recovery efficiency in existing fields), adding significantly to total supply.

Much of this will be driven by the unparalleled and largely unpredicted explosion in American oil or tight oil output.[2] But three other countries have immediate major potential from projects already in the pipeline and well under way. They are Iraq, Canada, and Brazil. Note how three of the big four on

2. A distinction must be made between shale oil and tight oil and oil shale. The more accurate term for shale oil is Light Tight Oil (LTO) – oil which can be extracted under hydraulic pressure mostly from shale rock formations. It should not be confused with oil shale, which is an entirely different product.

the increase are in the Western Hemisphere and only one in the 'traditional' Middle Eastern producer zone.

Only two current major oil producers are on the downward path. They are the UK (North Sea) and Mexico. But even these 'minus' areas may turn out differently. New deep-sea recovery technology may yet slow further North Sea decline, while Mexico's policy shift towards opening up its oil fields could halt the slide there.

American growth, thanks to its so-called 'shale play', has been on an incredible scale. The Bakken and Three Forks tight oil formation (in North Dakota and Montana) could on its own put the US on the Middle Eastern top producer scale, and there are some twenty other US formations of similar size.

All this could be profitably extracted. So long as the price of oil internationally stays in – or hovers around – the $40 to $70 range, production will go ahead. By 2020 the USA will have joined Saudi Arabia and Russia as one of the top world oil producers – all in the range of 10 to 12 million barrels a day – and leaving the rest of the oil-producing world trailing.

Meanwhile, two other giant oil producers of the past, but poor performers in recent years, could lumber back to centre stage. Iran in the 1990s was producing (and selling the majority of) some 6 million barrels of oil a day. Under sanctions and general underinvestment, this has sunk to just under 3 million barrels a day but could easily spring back as relations between Iran and the West thaw and oil sanctions are lifted. Japan, for example, which used to have big investments in Iranian oil and a big import take from Iran, will be only too ready to move back in with new money and new technology.

Iraq used to talk of raising production of oil from around 3 million barrels a day maximum under Saddam Hussein to 7 million and then 12 million by the 2020s. That big talk faded with the continuation of violence and extreme political instability, and almost vanished when the country looked as though it was about to be overrun by Daesh or Islamic State, the jihadist ultraviolent extremist movement which caught foreign experts (once again) so much by surprise by its rapid advance.

But with the barbaric Islamic caliphate now beginning to be contained, the trend will be reversed quickly, and export volumes, which had languished below 2.5 million barrels a day, could soar upwards – immediately to 4 or 5 million barrels, and in five years to 8 or 9 million barrels of exportable, high quality (sweet) oil.

Add to all this the even more dramatic 'gas' story, with natural gas production in the United States – again mostly from hydraulic fracturing and horizontal drilling (both well established technologies) – eliminating almost all gas imports and with the US beginning to be a supplier to world gas (LNG) markets, and the scene of abundance almost overwhelms. Cheaper oil means slower demand growth for gas at a time when supply has never grown more strongly. The outcome is obvious – higher supplies all round, flatter and slower-growing demand. A picture inevitably emerges of the world floating on a sea, a flood tide, of plentiful oil and gas.

Not only the United States but Australia and Canada as well are already becoming giant energy exporters (Australia now already being the world's biggest exporter of frozen natural gas). New fields in the Eastern Mediterranean await the accountant's and the investor's start signal when the economics become clearer, and when long-standing disputes, some diplomatic (between Cyprus and Turkey), and some over seabed boundaries, are finally settled.

Israel is already drawing large volumes of gas from its Tamar fields (and selling gas on to Egypt). Russia is clearly interested in being involved and may have associated plans to move its naval base from a collapsed Syria to a site in Cyprus.

Of course, over time, the usual commodity cycle asserts itself with high prices leading to low prices, investment deterred and in due course shortage, which means high prices again.

Conversely, once prices fall, the least efficient and highest-cost oil producers after a while close down, big new projects for oil expansion are postponed, and supply growth slows, in turn hardening prices again. Marginal production starts up again, projects are revived, and so the cycle continues. This will come, but at what pace and over what time frame nobody knows.

But this time, there may be a difference from past cycles. The biggest sources of 'oversupply' to the oil market and the world gas markets have clearly come from US shale producers. Some of these have been wiped out by a year or so of low oil prices and ultracheap gas, but unlike the oil giants, they can also come back into play as quickly as they have to leave. Shutting down and opening up when you are a small player is merely a matter of a few days' rearrangements and then switching on the pumps again. While the energy majors may take months or even years to resuscitate pulled projects, the small guys can start again as soon as there is a hint of price recovery.

Staggering improvements in the productivity of shale operations, driven by unending 'Yankee ingenuity', have further helped the shale operators stay in business while crude prices and gas prices drop.

While Shell and BP close down North Sea operations and postpone, maybe for years, vast projects such as Shell's gas-to-liquids scheme in Qatar, or further expansion of Canadian oil sands at Fort Murray in Alberta, the small guys pause and then leap in again at the first sign of price recovery.

The result? That prices stay constantly down for far longer than in the past, and production keeps returning persistently to price-killing levels. This may now be the supply pattern which will keep oil prices low for far longer than some producers assume or hope.

There is an additional factor – a misconception – which could work the same way. As was mentioned in Chapter Two, Arab leaders within OPEC have persistently argued that they are not going to cut OPEC production because 'it is up to non-OPEC producers to behave responsibly'. But how likely is it that, say, Russian oil magnates or American wildcatters are going to behave in any other way than responding to market forces, which means maximizing revenues at any price level at which they can afford to keep producing and selling?

A naïve but understandable presumption exists, especially in the Gulf, that Western countries, or producers such as Russia or Brazil, can command and control oil industry behaviour, checking output and export by diktat. They cannot.

The OPEC dilemma

I am meeting Suhail Al-Mazroui, the dynamic, young energy minister of the UAE. We sit in a circle of chairs surrounded by football fields of carpet and empty space in one of the many great halls of Abu Dhabi's Emirates Hotel. This is an edifice built on the scale of Imperial Rome, with machines dispensing gold bars in the main foyer. Down a wide, marble staircase in even bigger hallways, music is blaring out and a strange stick dance is being performed which I shall shortly be invited to join.

Crude prices have been falling like a stone and now stand at half the level they were a few months before. Minister Al-Mazroui is very clear. He and his fellow OPEC members will hold production levels (at around 30 million barrels a day – roughly a third of world output) and definitely not cut them. It is, he says, for other non-OPEC countries to 'show responsibility', clearly referring to the soaring output of America's shale producers as well as the unrestrained output from Russia and other sources.

But how wishful is his thinking? Is anybody in Russia minded to 'show responsibility' and are the activities of American wildcatters and thousands of small-time shale drillers under anybody's control at all? Arab leaders tend to assume that foreign governments control the spigots just as they assume that the British Government controls the BBC, and they indignantly complain when it airs criticisms of their societies and regimes.

The reality is that non-OPEC oil producers will carry on pumping. The only restraint will come when the higher-cost and less efficient operators among them simply cannot make any profit. Predictions that shale oil and gas operations in the US will be knocked out by a lower oil price are not working out. Some new drilling has been curbed and some smaller mom-and-pop wells have been shut down. But the bulk of the industry has pushed on, cutting costs and using new techniques to raise productivity. Any significant reduction in America's surging oil and gas output could be a long while ahead, and in the meantime, oversupply is bound to prevail, stocks will accumulate, oil storage onshore and in anchored tankers will brim over.

This in effect leaves the OPEC producers powerless. Speculation that the OPEC leaders are following a cunning strategy and keeping oil production high by choice are very wide of the mark. There is no choice and no alternative. Cuts in OPEC quotas would have little or no effect on crude oil prices – except possibly very briefly in the rumour mills of oil trading – and would simply deprive countries like Saudi Arabia or the UAE of both market share and revenues. This alternative 'strategy' does not exist. For the time being, there is no choice, only a reaction to developments far beyond any OPEC member's control.

Are there steps which the oil-rich Gulf States can take to escape from this price and production trap? Could there be some crisis or a series of shattering events which could push the oil price back to the happy $100-plus days and once more make simply exporting oil the way to riches? We can be sure that the wiser leaders amongst the low-cost Middle Eastern producers, who have had it all so good for so long, are thinking hard about different ways of using their oil and other commodities as the bases for new industries. The less wise will be clinging to the hope that 'something will turn up' to send oil prices soaring again to the levels of the past five years and put OPEC back in the driving seat. But they may also be wondering whether this time the luck and easy wealth which they have enjoyed over decades of good fortune may at last be running out.

There is the added factor than even within twelve-member OPEC, producers like Iraq, Iran, and Venezuela may simply ignore any quota discipline and keep the oil flowing as they desperately try to maintain revenues. As it loses influence, OPEC becomes a more quarrelsome body, challenging or simply refusing to understand the position of the dominant Saudi partner in the organization.

Venezuela, an OPEC member, is a good example of a country and a political regime that is unlikely to show the slightest restraint. Its very survival depends upon somehow keeping oil revenues flowing. Its policy of sending subsidized oil cargoes to neighbouring countries (such as Cuba) in the name of socialist solidarity has long been abandoned. Past schemes for repaying Chinese loans with big crude oil shipments, running to half a million barrels a day a year, have had to be revisited. This is a cornered country that is not going to cut production for anyone or in response to any argument about the need to behave responsibly and calm oil market volatility.

Oil Demand. It's not as simple as that!

On the demand side, there are equal uncertainties – and again some illusions about how strongly world demand will carry on growing. The absolutely central question, leaving every decision-maker and investor throughout the energy chain on tenterhooks, remains the same: is the world (and are worried finance and budget ministers) experiencing a short dip in world crude oil prices? Or is it time to prepare for a long era of low prices (compared with the immediate past)?

The conventionally reassuring view in the oil-producing world has all along been that while crude prices may dip from time to time, the long-term thirst for both oil and gas as energy sources will rise and rise, driven in particular by the two awakening giants – China and India – and by the rising needs of all the emerging markets of Asia, Africa, and Latin America as their new middle classes take off and consumer needs scramble the growth path followed by Western societies over the past century.

China, insist the oil optimists, will be putting 30 million additional cars on its roads every year, all needing oil. India, already a massive oil importer, will need millions more barrels per day. Economic growth and population growth, closely related as they are, can only mean more oil consumed – or so the reassuring chorus goes.

Yet there is something shaky in this confident oil industry view, something that does not add up. We are seeing once again here the familiar tendency of experts to read the future by extrapolating the past. There is no guarantee at all that Chinese economic progress will follow the same pattern as the West's. On the contrary, China has said loud and clear that it will not be following this route. 'Enough' has been proclaimed at the recent Fourth Plenum in November 2014 as the guiding principle of the next phase of China's development, put graphically in the adage that 'Chinese people need no more than three pairs of shoes'.

Not only will the 'car mania' of the twentieth-century Western world be discouraged in China but the whole pace of transportation technology will also move in a new direction. Even if official Chinese intentions are frustrated, the great engine of technological advance roars ahead – in some areas faster than ever – delivering vehicle and mobility systems propelled less and less by oil and petroleum products, using staggeringly little oil to cover large distances, or not requiring oil or petroleum at all.

Oil as a fuel for power generation is already anyway vanishing from the whole planet. The last bastions of oil for electric power are in the Middle Eastern oil-producing countries themselves, where wildly inflated fuel subsidies and massive use of heavy fuel oil for power stations combine to support a demand level which has vanished almost everywhere else, except in a few small states still unhappily dependent on imported diesel to run their outdated local power systems.

Japan has been a temporary oil and gas recidivist, returning to very large imports of both oil and LNG following the closure of all nuclear power stations which previously produced 35% of its power. But this is entirely temporary. The nuclear power stations are there. The capital – much the biggest part of nuclear electricity's cost – has been sunk. At least half of them will come back on stream now that the Japanese Prime Minister, Shinzo Abe, is well ensconced for several years of political power. A further heavy drop in world oil and gas demand will automatically follow.

This is a global future where of course there will still be oil demand – not least in industry, especially petrochemicals, where oil's unique properties make it by far the best and most competitive material. But it will not be as strong and it will not create the kind of demand pressures that have held up prices (with occasional dips) ever since the oil shocks of the last century.

Eating into the kingdom of oil, but so far only at the edges, will be those green alternatives which are either becoming cheaper through technology or are being supported by governments attempting to meet various carbon targets by penalizing fossil fuels as much as they dare and by subsidizing low-carbon substitutes, again as much as they dare.

The effect of these policies will be marginal because the costs are so high, falling on both consumers and taxpayers. Technology will do more than any amount of carbon pricing, emissions penalizing, or general taxing, to motivate consumers to turn to 'green' renewables. At the consumption end, smart grids and smart meters, hybrid, electric and low-petrol-burning, high-mileage vehicles, and a thousand and one other energy-efficient devices will do more than any government targets to moderate all forms of energy demand.

In fact, they have already done so. The new Western model is already alive and well and prospering. Energy demand generally, and oil demand in particular, have flattened completely in the USA and Europe. This is not just the effects of recession, holding back oil and gas demand, pending a new surge when recovery in a stagnant Europe takes off. No, we are looking at a structural shift to a more energy-efficient pattern of progress. The Chinese are going to follow suit. The flattened Western consumption of oil is going to become the flattened Asian consumption of oil. Demand is going to stay weak with or without a recovery in world economic expansion.

As the commentator and blogger Nick Butler puts it, 'The overall impact of innumerable technical advances is to limit demand while expanding supply. Add to this the renewed concern around the world about energy security, which leads inevitably to higher subsidized domestic production of at least one source of supply *and you have the recipe for a long-term decline in energy prices* (my italics).'[3]

The geopolitical consequences of this new world energy scene of high supply and low demand are enormous. Preparations are going to be required for a complete revision of strategies across many countries. Powers and hopes are going to trickle away in some areas and 'lift all boats' in others.

Amongst the petrostates – the countries which rely heavily, sometimes almost entirely, on oil and gas revenues – all lose to some degree, the states with higher production costs and less efficiency being hit the hardest.

Some measure of the immediate impact of sagging oil prices on various oil-producing states – and therefore often the impact on their entire political structures and stability – can be drawn from the break-even price for various different oil-producing provinces round the world – the crude oil price per barrel all the main OPEC countries need to balance their budgets. For Saudi Arabia and Russia, this is estimated to be $105. For Libya, it is $180, for Iran, $130 and Iraq, $100. Kuwait and Qatar can manage at $80. With the crude price nearer $40,

3. Nick Butler, *Financial Times,* 1 December 2014.

all are bleeding, all with political consequences, some yet to emerge.[4]

For more than forty years, the Organization of Petroleum Exporting States, OPEC, has dominated. Accounting collectively, at its peak of power, for 42% of world oil production through its twelve states, the leading countries – Saudi Arabia, Kuwait, and UAE – could dictate world prices, either by pushing them up through production cuts, or calming them down through announcing increased production.

This they did, acting in their first years as a tightly united monopoly or cartel pushing prices up to unheard-of levels, spreading panic through Western capitals – from $3.50 in the 1950s and 1960s up thirty times as high in the next decade. There they kept the price, stabilizing it most of the time by use of both production spurts and cuts – metamorphosing their self-proclaimed monopoly role into an oil market-calming body, acting, they insisted, in the universal interest by preventing extreme oil price volatility, but also keeping the crude oil price well up in the lucrative range to maximize revenues at around $100.

No longer. In 2014 'king' technology finally elbowed the OPEC powers aside. The writing was on the wall, and had been from around 2008, as advances in the already well-known methods for extracting gas from shale formations, and later from oil formations, began to be exploited extensively in the USA. Non-OPEC oil output began to soar. The Russian oil industry, decrepit for years after the fall of the Soviet Union, began to grow fast. Recovery rates from existing fields vastly improved with new techniques. All but blindfolded economic

4. Deutsche Bank Strategy document, July 2015.

'experts' could see the making of the oil and gas glut, the historic story of the industry – 'too much oil'.

All this would have weakened OPEC marginally but on the demand side, another steel pincer was closing in. As earlier described, a peak was coming and then passing, not in oil *output* but in oil *demand*. In the Western world, immense improvements in energy use efficiency were underway, reflected in the disappearance of gas-guzzling motor vehicles and the trebling of mileages per litre for conventional petrol engines. Natural gas, emitting 40% less carbon, was crowding in on oil and petroleum products not only in the transport sector but in many other fields as well.

And gas production itself was leaping upwards, producing in the USA the lowest price levels in modern times, taking America's large gas demands off the world market. Other gas projects round the world raced forward – for example, in the Caspian region, in Russia itself, in Norway, in Australia, and in the Eastern Mediterranean, led by Israel. The Israelis achieved a remarkable energy turnaround, moving in a few years from being reliant on Egypt for natural gas to the opposite – supplying natural gas *to* a struggling Egypt which had lost the will or capacity to develop its own energy resources. This may be reversed again as Egypt finds the will to develop its own substantial and recently revealed offshore gas, but for the moment, Israel has the upper hand.

And on top of all this, the internal OPEC discipline began to crack, as already described, with neither Iraq nor Iran nor Venezuela being prepared to agree to stick to quotas. The shakier the regimes, the more they needed high oil revenues for their survival (and in some cases, to keep their lives). No one was going to cut production (and therefore revenues) only to see others cash in.

What the Saudis found by 2014 – although not to their surprise – was that cutting production would no longer firm up the oil price. Other producers would simply step in and take the Saudi – or the Kuwaiti or the Emirati – market share. Cutting would mean losing, at least in the short term.

What goes down comes up

Eventually the trends will reverse. Eventually lower oil prices will knock out costlier production, delay or abort new projects, weaken new investment all round, and leave the producers with the lowest production costs – mainly the Gulf kingdoms and emirates in possession of the field. The price will then creep up again but much more slowly than in the past because the world's thirst has become much more moderated – again thanks mainly to technology. Government policies to restrain fossil fuel use anyway, in the attempt to curb carbon emissions growth, will also play a small part.

But it will all take a long time. Two years? Ten years? Who really knows? And in the meantime, heavy budget losses and massive reductions in state spending and patrimony will create political upheaval.

Iraq, which until 2010 had high hopes of mending its broken state through a big increase in oil exports and oil revenues, now finds that its resources for defeating the Islamic State are seriously depleted. Iran, which may have been anticipating a scenario of lifted sanctions following some kind of agreed curb on its nuclear ambitions, and the revival of its once massive 5 to 6 million barrels a day output levels, will now see those hopes postponed. (This will no doubt be to the satisfaction of Saudi Arabia.) Nigerian resources for tackling the increasingly ugly jihadist incursions from its north are being weakened. Mexican plans to liberate its oil sector will be frustrated.

Above all, the Russian Federation, whose oil and gas output was creating the monetary power base on which Vladimir Putin's international adventurism rested, now sees that base crumbling. The initiative now passes back into non-Russian hands, hopefully the right ones. The oil price has decided things. The outside powers – mainly the West – can either sit tight and watch while internal Kremlin struggles in due course unseat Putin and maybe grind Russia down into endless instability and social breakdown, or they can engage in a positive and helpful way to ease Russia through yet another spasm of tensions, travails, and violence.

The nations which are both big consumers and big producers of fossil fuels face a more mixed prospect. Lower oil prices for a prolonged period will undoubtedly check the shale gas and oil boom in the US and seriously weaken the economics of oil sands operations in Canada. Already the boom atmosphere in Fort Murray, the centre town of oil sands operations in Alberta, has given way to unease and gloom. Luxury properties stand empty. Unemployment is rising. Although much of the cost of oil sands production lies in the huge initial capital costs, and the marginal cost of each barrel is low, the apparent end of $100 oil has turned off new investors.

Estimates fly around about the dollar level at which investment in new shale drilling would dry up, but it is fairly clear that smaller drilling businesses will be in difficulties quite soon. 'Production from shale wells comes on fast and drops off fast', says Allen Gilmer, Drillinginfo's chief executive. 'So their economics are more exposed to short-term prices'.

Some will try going back to gas operations, but there, too, low prices have squeezed profitability for all but the lowest-cost operations in the easiest areas, such as Marcellus, Wolf Camp, and Eagle Ford. All in all, there will be a modest slowdown in the wilder side of the American shale phenomenon, but not a close-down. Furthermore, as Christof Rühl, former chief economist of BP, has pointed out, thousands of small-scale fracking operations in the USA, while temporarily knocked out by the oil price drop, can just as swiftly come back in as soon as there are signs of the price rising again. Thus, resilience and almost instantly rising production could constantly hold prices down again, lengthening the period considerably before market fundamentals eventually reassert themselves.

For Europe almost all fracking plans will now go on hold. In many countries they have anyway been politically blocked. In Britain, as described earlier, strong government enthusiasm has already been undermined by extraordinarily clumsy political handling and presentation. $40 crude prices, dragging down gas prices, will now put the stopper on most further investment in drilling – for the duration.

North Sea Travails

For Europe as a whole, the transfer of funds from the producers into the pockets of the consumers, and into reduced industrial costs, is proving a major 'shot in the arm' to compensate for its many other disasters and policy errors. The UK economy, already growing in contrast to the main Eurozone, is being further boosted, and while North Sea oil and gas operations have taken a hit, this comes at a time when the whole North Sea saga is entering its endgame. It is estimated that between 12 and 24 billion more barrels of oil lie beneath the UK Continental Shelf, and some gas. This would anyway have required major tax breaks and incentives to make it economic. Now these will cost much more and probably come more slowly. Lower oil revenues will hurt North Sea operations, and strong political pressures will be put on the Government to 'do something'.

Concessionary tax gestures have been made, and more will be needed, including more help with the expensive business of decommissioning huge but redundant North Sea structures and plugging exhausted oil fields. A newly empowered Oil and Gas Authority will have the task of nursing the whole North Sea industry through a difficult downsizing period. The development of the UK Continental Shelf stands as an incredible feat of British engineering and innovation, and a past source of enormous tax revenues (roughly £330 billion since 1975). It is certainly not to be abandoned but it obviously now requires a different approach.

However skillfully this is managed, it is impossible to 'buck the market'. A prolonged low oil price inevitably means North Sea field closures, big job losses, and an intense search by Britain's well developed oil supply industries for new customers overseas. But it also ensures that consumers feel richer every time they fill up at the pump and have more to spend elsewhere. In the end, although not before there has been some awkward and painful readjustment which sensitive public policies should seek to cushion if not resist, higher economic growth will far outweigh the disruption.

China, the world's biggest oil importer, will find cheaper oil a welcome lift to its economic advance, which had been slowing somewhat in 2014. Its own shale projects will go on hold and were anyway encountering numerous difficulties, but its 9 million barrels a day of imported oil, together with big gas imports, will suddenly come at a much lower cost. China's recent deal with Russia to take large gas volumes over a twenty-year period ahead, announced with a flourish in the summer of 2014, is no doubt being swiftly 'revisited' by Chinese authorities seeking certain price 'adjustments' to take account of new world price levels.

From the beginning of the oil industry, oil men have always had to cope with major fluctuations in the crude price. In 1859 it was $20 a barrel, and in 1861 it was 52 cents. Two years later it averaged $8.15, and in 1867 it was $2.40.[5]

Volatility is the norm, not stability. The long-past period of crude oil at around $100-110 is exceptional.

The dangers and disturbing geopolitical consequences have been listed in this chapter, and there are plenty more to come as it gradually sinks in to governments round the world that the high-level price plateau is a thing of the past.

Against this there are the undoubted benefits. The forecasting record for oil prices has all along been abysmal. This is because the determinants of price lie far outside the oil industry and far outside conventional economic analysis.

Is Lower Better?

A permanently lower range for oil prices – if that is what is in store – has many advantages, even for producers.
As J. D. Rockefeller put it long ago:

> 'However great the fun and the profits of making oil very dear, in the long run it does not pay; it weakens markets and stimulates competition.'

A much lower price than the $100 level of recent years trickles through into lower industrial costs everywhere. Smaller nations without their own resources – especially small island states – find their import bills miraculously reduced. Governments in oil-producing states ease up on their relentless revenue squeeze on

5. Paper by Dr. Carole Nakhle.

the energy industries. The 'fiscal storm' quietens. And as is always the case in reasonably free economic conditions, fascinating new opportunities for new ventures and new businesses open up. Technologies as yet unseen open up new vistas. Ideas and disciplines which lived apart come together. The human network of information and exchange carries on being woven, making possible the previously impossible.

The oil and gas industry has an amazing track record of survival and reinvention at almost any oil price level. Not only will the core oil industry business remain, it will remain fundamentally sound and profitable also.

All this is in total contrast to the perception of oil as a fading feature of the global economy, and to predictions (and hopes) that we have reached the end of the oil age. The policies and the strategies based on this latter assumption are all going to be disappointed. If the world is serious, rather than just emotional, about combatting climate change and reducing carbon emissions, it will have to proceed on entirely different tracks to those now being followed.

Two major questions now swing into primary position in the whole cosmos of energy supply. Can the low-carbon and green alternatives, especially nuclear power, ever deliver on the scale and at the price which the world will require as modernization spreads across the entire planet? And what path will China, by far the most energy hungry entity of all, follow as it races into the high-growth and high-consumption age? To these issues the next three chapters turn.

CHAPTER TWELVE

But When Will Green Mean Cheaper?

The price prospects for renewables – how much dare governments charge consumers? – when can renewable energy sources become commercial and competitive? – the encouraging American experience – research and the Global Apollo Programme.

The cost of some low-carbon renewable sources of power could be about to fall sharply. Some, but not all. The biggest low-carbon power source of all, nuclear, has shown almost no signs of being on the downward cost path. The bills for new nuclear stations have risen and risen. This is so disappointing, and so central to low-carbon prospects worldwide that we will reserve the nuclear story for the next chapter by itself. Here we survey the quite promising outlook for a range of green and low-carbon energy patterns, some already in reach, some nearly so, and some needing the extra push of research and development which is where green-promoting funds ought, of course, to be going.

As of now, electric power from most low-carbon renewable sources, whether wind, sun, or nuclear power, is still desperately expensive, in all except very local and very small-scale conditions. This is what has defeated most alternatives to fossil fuel again and again, and this is what has led governments to attempt so many interventions to discourage fossil use and incentivize renewables investment – mostly with disastrous and counterproductive results.

The one big exception is hydropower. Hydropower-driven turbines now generate about 15% of the world's total power. Within the OECD countries, this is probably near the peak – there are no more good mountain and river sites and there is anyway immense environmental objection and resistance to any new projects. But in Asia, Africa, and Latin America, it is a different story. The World Energy Council estimates that only 33% of hydropower potential has been developed in Latin America, 22% of potential, in Asia, and only 7%, in Africa.

Hydropower remains on the competitive side of the cost curve for generating electricity, although of course delays and environmental resistance could keep raising the cost barrier and deterring further investment. But for the other renewables – and wind power in particular – the picture is quite different.

The highly political questions are these: how heavy a burden dare governments lay on consumers and taxpayers to meet the extra costs – the so-called externalities (CO_2 in particular) – and for how long? What are they being asked to buy, and is it worth it?

The cost-benefit tally list goes something like this:

Extra charges, over and above basic power needs, on households and on industry bring a lower-carbon world, a cleaner environment, greater security of supply, and eventually lower energy bills, as the cost of renewables comes down with scale, maturity, and technological advance.

Climate campaigners like to argue that payments are in effect an insurance premium against the risk of dangerous global warming. But as soon as risk is mentioned, it opens up a debate about how much risk and the value for money represented by the insurance premium. Some also argue that if it is future generations that are being protected, then they are the ones who should pay.

The climate lobby also argue strongly that extra payments now will bring lower prices later, as the green technologies find their feet and costs are reduced.

Of course, these extra charges are also buying something not so attractive. They are buying forests of wind pylons across the countryside, which are mostly but not always environmentally offensive. They are buying biomass processes which may be far from carbon-reducing and they are buying into a redistributive pattern which is generally perceived to be unfair, since consumers' money may be going not just to new low-carbon investment but to substantial profits and benefits to corporations and landowners, including China General Nuclear Corporation, which has purchased large stakes in UK wind farms (detailed in Chapter Fourteen).

The balance in all these arguments has to be political. It cannot be objective because while both sides claim they have 'the facts', they actually have 'the beliefs'.

The balance suggests that a premium of a certain size can be charged, but the matter of how much is pure politics. It varies from country to country, period to period, and from energy type to energy type. The US Department of Energy gives one estimate – that electricity from onshore wind farms is 17-18%

more expensive than electricity from gas turbines. But this figure should be treated with caution. It is not clear whether it takes into account system costs and the 'intermittency' problem, which requires all wind installations to have backup from other parts of the grid, usually gas-fired facilities that themselves require subsidy to get built, since they stand idle when the wind is blowing.

Countless other factors far removed from energy and climate issues also come into the calculations.

For example, if a population already feels overtaxed, that will limit what will be tolerated. If the climate arguments are made weakly or are skillfully challenged, that, too, will undermine the 'saleability' of higher green charges. If charges are imposed without adequate explanation and discussion or in a spirit of arrogance and impatience, that, too, will produce reaction and rejection. Something of this kind has certainly occurred in the British context, as we saw in Chapter Nine.

Appeals to secure supply and self-sufficiency may also lack pulling power in some societies. In the British case, self-sufficiency in the past proved a very mixed blessing. There is a strong folk memory that in the last century, when the country thought it had secured home-grown power supplies in the form of coal, it found to its cost that the security was not there. The coal miners went on strike and the nation was brought to its knees. Blackouts ensued and industry went on to 'a three-day week'. More diversity of imported sources might well have been safer.

Who Pays – and for What?

The issue comes down to costs. If we are now moving into a longish era of cheaper fossil fuels, the cost of low-carbon will be higher still. Estimates vary widely and erratically, with pro-green lobbies giving one figure and climate sceptics, another. But the median figure in Britain (which may be roughly similar elsewhere in Europe) is roughly an extra 40% on every energy bill by 2020.

Can a government justify or 'get away with' that level of impost and still survive? Possibly yes, if it is believed that the extra premium is not only buying protection from violent climate events within a foreseeable timescale, but also protection against the vagaries and dangers of fossil fuel supply – and also if the longer-term costs of renewable sources are seen to be genuinely moving downwards. The infant industry argument – that industries need early protection to allow them to achieve scale and cut costs – does have some justification if the infant is going to mature into an adult performer. But the 'infants' to be protected have to be picked with caution and care, and the support funding, strictly time-limited.

If oil and gas supply sources are seen as politically unstable (which many of them are), if transmission and delivery systems are seen as internationally fragile and vulnerable, then the case for paying up for greater security from other sources grows stronger. But these points need to be put with a persuasiveness and clarity which has been sorely lacking, especially in Europe, where both families and industries have been asked to carry the biggest extra cost burden, with pathetically scant explanation beyond sweeping generalizations about saving the planet.

Technology may or may not deliver the cost reductions needed. For the incurably intermittent sources – notably wind power – probably the most hopeful way forward is via the storage technology. If storage battery systems can be produced which are both cheap and effective, a whole slice of costs for the most intermittent alternative sources – notably power from wind when it blows – is removed. The need for extensive backup generators, invariably and unavoidable fossil fuel-powered, would disappear.

Cost and size are the barriers here. At Sanyo outside Tokyo, I was shown the latest in storage ideas. It is a building about 50 metres long and two stories high. It is crammed with lithium batteries. Fed with power, it will hold about 1 megawatt for six months before it all seeps away. The cost of 'delayed' electricity via this route is astronomical, about five times the cost of direct flowing current, and knocking all electricity stored this way out of the competitive frame by miles.

A Norwegian firm, NEST, believes that it can do storage much cheaper. Its plan is to heat up small concrete blocks made of special materials to 66 °C. It believes it can be held there for at least six months and then released as steam power. The claim is that it works, but somehow no one is yet buying it.

Another big hope for lowering the costs of decarbonized electricity comes from the coal sector. The dream is that carbon can be sequestered from the fumes of coal-burning power stations and piped away somewhere using the technique of carbon capture and storage (CCS), on which earlier chapters have already touched. The idea is simple but its implementation is fiendishly difficult – and expensive. It involves adding new machinery to existing plant, or purpose-building new coal-burning plant with siphoning devices which extract the carbon from other emissions. Sulphur is already extracted from most coal-burning stations, so why not carbon? Very long pipelines are required to take it away – but to where?

The answer has to be to exhausted oil and gas fields far out in the North Atlantic. Miles of piping are needed, all at a cost which turns coal, the cheapest power source for electricity, into much the most expensive. With the present state of the technology, it cannot be done. Enormous sums have been poured into experimental projects both by governments and by the big energy companies. In 2009 the UK Government earmarked a staggering £4 billion to be spent over a number of years on CCS projects. Five years later, nothing worthwhile had emerged. Some projects were abandoned as the costs kept rising; some staggered on.

The determined carbon reducers keep their spirits up as they watch coal-burning round the world grow remorselessly, rather than shrink, with the hope that new technologies can be found to make the whole process much cheaper. Perhaps the carbon sequestered can be processed on site. Perhaps it can be turned into limestone slabs and trucked away.

None of these projects have so far delivered. One in the UK is going ahead at Deeside. A bigger one at Peterhead was cancelled – too costly to go on being subsidized. Round the world there are small, experimental schemes for applying CCS to fossil fuel power stations, about twelve in all with a further ten planned. The Algerians have one for burning oil. The Australians have one for coal. The Canadians, another. All new power stations being built are required in these countries to be 'CCS-ready,'; that is, the land space must be there for the full-blown CCS apparatus which will arrive – one day.

Green Breakthroughs

Big breakthroughs in electricity storage methods at dramatically lower cost, and carbon capture systems also at much lower cost, would be the making of large-scale greener energies. Instead of being two or three times as expensive, as well as more unreliable, electricity through these would be actually cheaper. In some areas in the US, this is beginning to happen. Aided both by remarkable new technology and by ingenious new financing methods, a solid economic case for investing in renewable resources is emerging, without the need for the crutch of special tax breaks or direct incentives and purchase obligations. Yield cos, crowdfunding, and securitization can all play their part. Electricity from wind, intermittent as it is and requiring heavy support costs both in infrastructure and integration into the grid system, could begin to be within striking distance of competing with gas turbines.

When this crucial crossover begins to be into view in Europe, this will be the moment when consumers will turn willingly to low-carbon sources, instead of being penalized, taxed, and hustled into doing so. The renewables industries will bring their prices down, instead of put everybody else's up. The whole energy landscape will at last shift towards safer, greener energy sources, happily, eagerly, and fast, driven by direct experience of paying less for the same power. That *would* be the day. It may come – but not yet.

One initiative which may help the push to cheaper green energy is the Global Apollo Programme which has a number of private backers and aims to focus research, on a Manhattan Project-scale, on the central task of bringing renewable costs down, making them cheaper than fossil fuels, and thereby harnessing the decisive power of the market and the consumer for the green cause. This makes complete sense if it can produce results. It is also of course what governments and public authorities should be doing, instead of pouring billions into subsidies of existing uneconomic projects which can only live off a constant government drip-feed.

Much nearer real commercial viability are schemes for replacing conventional domestic boilers with heat pumps or boilers with solar thermal or passive flue gas heat recovery. In millions of homes, not just in newly built ones, the familiar gas boiler may be a thing of the past. Decarbonizing domestic heat – which in the UK produces 23% of emissions – may actually be made to pay off. A scheme drawing heat from Thames water, pioneered by Mitsubishi Electric and delivering domestic, low-cost heat to hundreds of flats in a new development, called Ecodan, may also be a model ready for wider adoption.

But – and it is a big but – for the homeowner, the pay-off has got to be large and immediate. It has to compensate for the usual 'horror' of building works in one's home, and it has to be free of large, upfront capital outlays – which most people do not have to hand. There is something almost laughably bureaucratic about schemes which involve taking out loans or putting up precious capital in return for the assurance that over the years lower energy bills will make up the difference or pay back the loan. It all has to be much more tangible and immediate than this. When newly designed heat pump systems, fully installed, are cheaper than buying new, conventional ones, then they will be bought. When solar panels on the roof are hardly much more expensive than a new TV set – and this may be the case soon – then they, too, will be installed.

But it is the really big low-carbon projects of the future which carry the greatest promise of the green world at least holding its own against fossil fuels. When Bill Gates announces that he wants to back low-carbon energy, where he is looking – and where governments should be looking – is towards the research and development which can in turn bring green power to whole societies and to billions of people.

Gates is said to be interested in smaller 'travelling wave' nuclear reactors, which produce little or zero waste and cut out the massive costs – and political worries – of lethal waste storage and disposal. He may also plan to back solar chemical power and high-altitude wind power, which could catch the permanent and steady solar winds. Like super batteries, which will be able to store and feed into the grid full high-voltage electricity when needed, these are the keys to unlocking a greener world.

But by far the biggest ally of all in the green cause would be the one which is struggling least successfully to bring down costs, which is furthest from commercial viability, and which is anyway least loved by the world's climate campaigners. This is civil nuclear power, always on the edge of renaissance but always facing setbacks. Can it ever break through?

CHAPTER THIRTEEN

Biting The Atom: To Build Or Not To Build

Nuclear power stations: the endless loop of doubt and decision – the lessons of Fukushima – facts about worldwide nuclear power station construction – is there a renaissance? – Germany and Japan go opposite ways – Hinkley, Hualong, and the Chinese nuclear entrée into Europe – is Hinkley C the last of the line? – prospects for small modular reactors – a race against time.

March 3rd 2011. 3pm. It was at this moment that the key staff at the vast Fukushima Daichi nuclear power station realized they were in deep trouble. The tsunami had not only cut off all outside power but had also surged over the station's defences and flooded out the diesel backup generators. There was no electricity to pump coolant through the reactors, and the temperature in reactor one was rising dangerously.

Without circulating water, overheating could lead to a hydrogen explosion inside the reactor vessel and maybe to full meltdown. At the very least, this reactor, and maybe the other two operating reactors, two and three, out of six at the plant, needed venting, meaning that plumes of highly radioactive vapour were allowed to be released skyward.

The official view had been that it could not happen. The station's defences against an earthquake-triggered tsunami wave were 13 metres high, and there had never been a tsunami that big in history.

But the word 'never' was wrong. Seven years before, the operating company, TEPCO (Tokyo Electric Power Company), had commissioned a study of past tsunami events in the area. The study had shown that over 1,100 years before there had indeed been a tsunami with waves higher than 13 metres. Soil and rock tests showed that at the time of what was called the great Jogan earthquake of 869 AD, waves much higher had swept inland at precisely the point where Fukushima was sited.

The study had been shelved and at least one of the academic experts involved in it had been asked to leave. It was unnecessary scaremongering, said company managers; it would damage public support in Japan for nuclear power (which was always uneasy at the best of times). There could never be a repeat. But there was.

The tsunami that day was one of the costliest natural disasters in history, with the cost put at $645 billion, with 15,870 people dead, and a further 2,846 still missing. 130,000 buildings were flattened and, as a result of the Fukushima crisis, land for 1,800 hundred square kilometres around, including villages and towns, was evacuated and remains uninhabitable to this day, and maybe for years into the future, contaminated with unsafe doses of radiation, mostly caesium. The technology to clean up the area does not exist.

Yet the events which followed at Fukushima, with the eventual abandonment and 'cold shutdown' of the plant, with the release of heavy emissions of radioactive materials, and with the heroism of staff working to prevent pressures and melting reactor rods from leading to a full nuclear meltdown, have had a curiously selective impact on the whole global nuclear industry.

Why was that? A different outcome might have been expected. But at root, while the Fukushima affair had ugly nuclear consequences, the cause was not nuclear, as James Lovelock reminds us in his most recent work.[1] It was a natural disaster of monumental proportions. It was a dreadful warning not against nuclear power but against the siting of nuclear power stations.

The nuclear power industry has had consistently bad press for half a century and has raised deep and persistent popular fears – especially in Europe but also in the United States. This is in contrast to the early days of civil nuclear power when the industry was greeted as the source of unending supplies of cheap electricity for all and for all time.

Sometime in the 1970s, the image began to slip, with the Three Mile Island incident in 1979 hastening speeding it on its way down, and with Chernobyl proving what many people believed would be the nail in the coffin of the whole industry.

But they were wrong. In the decade after Chernobyl, a dozen countries, mostly in Europe, decided to phase out their nuclear programmes and build no more. But the two big European builders – France and Germany – remained completely committed, with France pushing ahead even faster with its remarkable civil nuclear performance – putting the finishing

1. James Lovelock, *A Rough Ride to The Future* (Allen Lane, 2014).

touches to its already built programme of no less than fifty-eight new nuclear stations, providing 70% of all electricity from nuclear power. As an engineering feat and a political exercise of enormous dexterity and skill, it remains unequalled.

The table below gives the worldwide position now – today – on nuclear power and the nuclear outlook, and the underlying facts.

SOME FACTS

The first commercial nuclear power plants entered operation in the early 1960s.

In 2012 nuclear power generation suffered a second year of record decline (-6.9%, or 177 terrawatts). Outside the OECD, nuclear generation continued to grow, albeit at a slower pace (3.7% vs. 5.9% in 2011). Globally, China (12.5%, or 11 terrawatts) and Russia (2.7%, or 5 terrawatts) recorded the biggest increments.[2]

Global subsidies to renewables reached $101 billion in 2012, up 11% on 2011, and need to expand to $220 billion in 2035 to achieve desirable targets of carbon reductions.[3]

The UK has around 100 gigawatts of installed electricity generation capacity – 40% comprises installations using gas, and 30%, installations using coal. About 8.1 gigawatts of current capacity is scheduled to close by 2020, including 3.9 gigawatts of nuclear-produced electricity. By the end of 2023, all but one of the existing nuclear power stations (i.e. Sizewell B) are due to close. About 4.2 gigawatts of mainly coal-produced electricity is due to retire by 2015.[4]

---

2. *BP Statistical Review of World Energy*, 2013.
3. IEA, 2013.
4. European Commission, 2014.

Country	Operating units	Under construction	Share of electricity	Share of World	Country	Operating units	Under construction	Share of electricity	Share of World
USA	104	1	19%	32.7%	**Japan**	50	2	2.1%	0.7%
France	58	1	74.8%	17.2%	**Brazil**	2	1	3.1%	0.6%
Russia	33	2	53.8%	7.2%	**Bulgaria**	2	-	31.6%	0.6%
South Korea	23	4	30.4%	6.1%	**Hungary**	4	-	45.9%	0.6%
Germany	9	-	16.1%	4%	**Slovakia**	4	2	53.8%	0.6%
China	17	29	2%	3.9%	**South Africa**	2	-	36%	0.6%
Canada	19	-	15.3%	3.9%	**Romania**	2	-	19.4%	0.5%
Ukraine	15	2	46.2%	3.6%	**Mexico**	2	-	4.7%	0.4%
UK	16	-	18.1%	2.8%	**Argentina**	2	1	4.7%	0.3%
Sweden	10	-	38.1%	2.6%	**Iran**	1	-	0.5%	0.2%
Spain	8	-	20.5%	2.5%	**Netherlands**	1		4.4%	0.2%
Belgium	7	-	51%	1.6%	**Pakistan**	3	2	5.3%	0.2%
India	20	7	3.6%	1.3%	**Slovenia**	1	-	-	0.2%
Czech Rep.	6	-	35.3%	1.2%	**Armenia**	1	-	26.6%	0.1%
Switzerland	5	-	35.9%	1%	**UAE**	-	1	-	-
Finland	4	1	32.6%	0.9%					
					World	**437**	**66**		**11.3%**

(IAEA, & World Nuclear Association, 2013)*

**The IAEA excludes Taiwan, which has 6 nuclear plants and 2 under construction*

The UK's Hinkley Point C, if it goes forward, is expected to save 9 million tonnes of carbon emissions each year – the equivalent of 2 million cars.[5]

There are over 430 commercial nuclear power reactors operable in thirty-one countries with over 370,000 megawatts of total capacity.[6]

China has completed construction and commenced operation of seventeen new nuclear power reactors between 2002 and 2013.

Is there a nuclear renaissance?

These tables and summaries portray a busy picture around the civil nuclear world, not a decline. In Britain – the pioneers of civil nuclear power – the first wave of nuclear building ground to a halt with intense interprofessional rivalries over the best reactor technologies to deploy for the first 'replacement and advance' phase. An attempt to end the bickering and restart the new-build programme was made in 1979 with the Government in the UK announcing its support for nine new reactors – in emulation of the French achievement – all of the Framatome pressurized water reactor design.

But this 'renaissance' ran into two major obstacles. The most visible was the rise of political concern and opposition following Chernobyl. But the even greater obstacle was cost. With oil and gas prices entering a phase of actual decline, the demands for expensive, new safety features rising, and political and planning resistance prolonging construction times and hence expense, the whole British programme began to slide away. In the end, only one new station was built – Sizewell B in Suffolk.

5. EDF, 2013.

6. World Nuclear Association, 2014.

The British scientific community came out of this fiasco poorly. Too many of its tyros simply forgot the cost factor (rather as today, too many low-carbon planners keep overlooking cost). Ministers were advised on the wondrously low cost of electricity produced from nuclear plants – said to be in 1979 a penny ha'penny a kilowatt hour – without any account being taken of all capital costs including the enormous costs of decommissioning plants at the end of their lives.

Only when the plan to privatize the electricity-generating industry emerged in the mid-1980s did sharp questioning from would-be private sector investors winkle out the full, costly truth. Nuclear power might have many advantages, such as security of supply on a full industrial-load scale, plentiful fuel supplies, and, as it came to be perceived later, very low carbon emissions. But it was not cheap power at all. In fact, it was very, *very* expensive, with the expense rising all the time as new safety requirements were added to construction specifications and building times lengthened at huge costs because of planning delays (many of them driven by intense political opposition).

Elsewhere in the world, new building pressed ahead in the 1980s and 1990s with China leading the way. The Chinese embarked on a colossal, new civil nuclear programme in the late 1980s. India began building seven new reactors. South-East Asia – Indonesia, Malaysia, Thailand, Vietnam – started on twenty-nine new nuclear power stations. A de-Sovietized Russia began again with new designs and new confidence, including work on the world's first floating nuclear power plant. Korea began growing a highly successful new-build capacity, constructing six dual reactor stations in Korea itself and winning contracts to build three more, including the giant new plant in Abu Dhabi – at 5 gigawatts, the biggest in the world. Even in the United States, where political opposition had been most vocal, plans began to be drawn up for two new plants – in Georgia and South Carolina.

By the first decade of the present century, expansion continued. With the exception of the handful of phasers-out and waverers, new construction raced ahead, with talk – once more – of a new nuclear renaissance. This time the industry had the uneasy and often highly qualified support of some environmentalists who welcomed the low-carbon potential from nuclear electricity. James Lovelock, the inspirational creator of the Gaia concept, shocked many green campaigners by converting to nuclear power as the one sure way forward for reconciling the world's power needs with low-carbon sources.[7]

This, then, was the scene when the March 2011 tsunami struck the Fukushima Daichi plant. Where has the tsunami left the global nuclear scene? Answer: caught on a wave of doubt with varying reactions in different countries and societies, facing, in many cases, different problems and challenges. The two most significant consequences have been in Germany and in Japan itself.

The key dividing issue is whether it is a matter of restarting nuclear power plants that have already been built, potentially have a long life ahead of them, and have performed well in the past (as in Japan and in Germany, for example), or whether it is a question of building brand new ones or new replacements (as in the UK, Poland, China, and India, amongst others).

In the first category Japan and Germany have gone opposite ways.

In Germany Mrs. Angela Merkel's Coalition Government quite suddenly decided to put its civil nuclear policy in reverse. In an extraordinary turnaround, Mrs. Merkel announced that

7. James Lovelock, *A Rough Ride to the Future* (Allen Lane, 2014).

all nuclear power would be phased out by 2022. A number of nuclear stations would be closed forthwith. Germany would go predominantly green, with power from Baltic wind farms and solar panels. This would be the great 'Energiewende', a vast transition from both nuclear and fossil fuels to wind and solar power.

The outcome of such an impulsive, short-term, and politically inept policy is proving entirely predictable, as was described in Chapter Eight. Imported coal, French nuclear power, and Russian gas have been called in to fill the gap. The availability of cheap coal from the US, denied a market there by the expansion of shale gas, has prompted a surge of German investment in new coal-fired plants and brought old coal-fired plants out of mothballs. Lignite (brown coal) has been brought in from the Czech Republic and Slovakia and now provides no less than 26% of Germany's electricity needs. Power costs have risen sharply – the only short-term curb being the widespread use of coal for electricity at costs seven or eight times less than the costs of power from burning natural gas.

Germany's biggest need for electric power tends to be in the south. But the hoped-for wind power, besides being intermittent, is in the north, and the power lines to bring it south have not been built. Popular resistance against big, new transmission pylons has imposed continuing delays.

In consequence, Germany has ceased to reduce its CO_2 emissions. Instead, in 2012 and 2013 they rose.[8] It would be difficult to think of a more perverse policy (although, as we shall see, there is one to hand).

The Japanese situation is different. Before Fukushima, Japan was producing 38% of its electric power from nuclear stations.

---

8. *International Herald Tribune*, 10 October 2013.

The plan was to increase it to 58%. After Fukushima, all plants were closed. This immediately created the need for vastly increased oil, natural gas, and coal imports, estimated to cost an annual ¥3 trillion (£30 billion) in added import charges.

The declared aim of the new 2013 Japanese Government, led by Prime Minister Shinzo Abe and operating from a position of unusual political strength, was to open at least half the closed nuclear power stations, if necessary sequentially, and cut the crippling gas and oil import bill. In the meantime, Japan would have to scour the world for reliable sources of gas and oil. This was the message from Mr. Abe when he visited the United States and surveyed its shale gas plenty, and when he visited Canada and discussed ideas for shipping its ample gas resources from Alberta across British Columbia to a major new export terminal at Prince Rupert.

But for all these arrangements, the question is: how long? And therefore, what kind of investment to meet what length of need? The obvious ideal would be to get back Japan's nuclear power as fast as possible and stop the haemorrhaging of cash. But despite having large parliamentary majorities in both the Diet and the House of Councillors (the lower and upper houses), until recently the political leeway to start reopening nuclear stations has just not been there. This may be partly because the Japanese electricity consumer has not yet felt the full weight of the extra fuel import costs. Highly successful energy saving was already cutting energy bills before Fukushima, and the non-nuclear plants had considerable fuel reserves. Once these have been used up and prices start rising, it will become painfully apparent what heavy costs are entailed by the continued denial of nuclear power.

Now that the Japanese Prime Minister, Mr. Shinzo Abe, has skillfully entrenched his ascendancy with a further 'snap' election giving him an even bigger majority, his political power to push for gradual nuclear reopenings is at its strongest. Events may be slowly coming together to enable Japan to restart at least half its nuclear stations.

But meanwhile, the situation has hardly been made easier by the actions, or inaction, of the Fukushima owners, TEPCO. Huge tanks storing thousands of litres of water used for cooling at Fukushima, which was possibly partly radioactive, have been allowed to leak into the nearby sea, and some into the ground. A further wave of fears about nuclear power, and further political resistance to a nuclear restart, were the consequences.

Eventually, and maybe by the time this book is being read, the Japanese nuclear industry will start up again, at least partially. Most recently, in April 2015, a Japanese court in Kagoshima has gingerly given permission for two nuclear stations to be reopened, although elsewhere in Fukui Prefecture, another court has made the opposite ruling. Progress will be slow. Its reactors are well designed and efficient. But too many of them are old and sited dubiously on earthquake fault lines at vulnerable coastal points. About half the existing plants could in due course be reopened and two more are under construction. Six more are planned. In the meantime, Japan's call on global gas and oil resources, which once looked ready to shrink dramatically, will instead stay extremely high – with significant consequences for global oil and gas markets. With oil prices having fallen by almost half in the last months of 2014, it needs no imagination to guess what will happen if Japan cuts its imported oil and gas needs and goes back to nuclear. What has dropped will drop further. World demand for oil and gas will ease further. Thus, the strange and complex linkages between all aspects of the energy mosaic are once again demonstrated. From Japanese nuclear engineers and political managers, the threads lead straight to global crude oil prices.

Asian Nuclear Technology

Meanwhile, the German giants, E.ON and RWE, have drawn in their horns, cut their overseas commitments and interests, including participation in Britain's faltering nuclear revival, and turned to other power interests. The Japanese have taken a different route. Their major nuclear construction industries, such as Hitachi, Toshiba, Mitsubishi Heavy Industries, and others, have increasingly looked overseas and are set to be major supporters of British nuclear revival plans.

Something of an international 'teaming up' strategy has begun to take place between the Pacific powers, with Japanese, Korean, and Taiwanese nuclear builders joining forces and with even some signs of Chinese cooperation. What we now see throughout Asia is a massive new-build programme which Japan itself will eventually have to rejoin. China is using its rapidly accumulated nuclear expertise and experience to probe out into the much wider world. Big efforts are being made to consolidate the three main and highly competitive nuclear builders. Eight new reactors were completed in China in 2015 with a further twenty-six under construction. Six reactors have been announced for construction in Pakistan and one – the Hualong One model – is planned for Argentina. Strong Chinese interest has been registered in the great Hinkley Point C project in the UK. With disarming candour, the Chinese have stated that their interest in this EDF-led project is to enable them 'to learn'.

If this costly project finally goes ahead, a lot will indeed be learnt by the participating parties. EDF and its subcontractor supplier, Areva – with which EDF has deteriorating relations – will expand its understanding of how to build stations of this type which actually proceed on time, within budget, and eventually work (an achievement not so far reached by EDF with its other recent projects at Olkiluoto in Finland, at Flamanville in northern France, and at Taichan in China, although the situation at the last is said to be a little better). The station at Flamanville in Normandy is seven years behind and €7 billion over budget. The Finnish project at Olkiluoto is eight years behind with the original budget now doubled. No plant of the kind that is proposed for Hinkley Point – the European pressurized reactor – is in operation anywhere in the world.

Latterly, the Chinese have grown much more cautious as they survey the shaky progress, extended budgets, and delays encountered by EDF in its other similar projects. The plan to overcome their hesitation is to keep the two junior Chinese partners in the Hinkley project happily involved and, in addition, to offer them the chance to take part in, and eventually control, two further UK nuclear projects, one at Sizewell in Suffolk and the other at Bradwell in Essex. Deals have been announced with a fanfare during the state visit of the Chinese President, Xi Jinping, to Britain in October 2015.

Two sweeteners were proffered to the Chinese to seal them in. They will be given a share in a second plant on a similar Hinkley model (the EPR design), and they will be permitted to build a further reactor plant at Bradwell in Essex using their own Hualong One technology. For the Chinese, these are both golden prizes, potential gateways to contracts for similar machines in other Western countries. As one senior Chinese executive put it, 'This is a victory for China'.

Yet the whole history of major nuclear plants shows that projects announced at a political level can remain a long way from projects physically going ahead. Further delays are bound to arise with the final go-ahead still many months away and construction workers still being laid off at the Hinkley Point site. The completion date for Hinkley, originally intended to produce electricity by Christmas 2017, then by 2019, and then by 2023, remains postponed further into the uncertain future.

Queries also remain about the safety of the reactor vessel design offered by EDF's supplier, Areva (the company formed out of the old Framatome enterprise and another smaller nuclear company, Cogema). The French Government may be facing an enormous bill – up to €5 billion – to inject capital into the ailing Areva company and to enable it to merge with EDF. All in all, Hinkley C is looking a very fragile and sick endeavour – a poor flagship for the UK's re-entry into the modern age of nuclear power.

No one really knows how much it will eventually cost to build, or how much of the gargantuan sums mentioned will fall on the British Government. First, the capital cost was to be £14 billion, then £16 billion. Later, much larger figures began to be mentioned – one even touched £25 billion. But this figure was scaled back to £18 billion when the Chinese finally signed up. Payment for the 'education' of the French and Chinese interests will be borne by the British energy consumer and the British taxpayer. The one party in the enterprise which has not learnt is the British governmental team which negotiated the whole, costly Hinkley Point project.

For future new-build programmes, the key is going to be cost, cost, cost – or at least it will increasingly be so in open societies. In more closed societies with less or non-democratic regimes, it will still remain possible to disguise or hide the immense subsidies required to build new, large nuclear plants, to cope with local protests, to handle waste problems, and to postpone or simply not discuss longer-term decommissioning costs. In China, in the UAE, in Russia, for example, this remains the position. No transparency and no detailed accountability mean that big and long-term public expenditure commitments to support nuclear new-build can be undertaken by inner political groups, maybe arguing amongst themselves but certainly not exposing the issues to open public debate.

This of course used to be the case in the UK as well, and explains how a whole fleet of early nuclear stations came to be built in a more deferential and much less transparent age. In a nationalized mammoth like the old Central Electricity Generating Board, costs and cross subsidies could be lost, and were, in a maze of half-revealed accounts.

A Giant Swiss Watch

Much older nuclear power stations ordered decades before were just then coming on stream. Just after its opening, I was invited to visit the advanced gas-cooled reactor at Dungeness, Dungeness B, which had taken no less than fourteen years to build. This was early 1980. Of course, all track of ever-rising costs had long been lost or covered up in mazes of figures within the accounts of the owner – the Central Electricity Generating Board – publicly owned and therefore owned by nobody. This was rather similar to the methods used by Electricité de France in financing fifty-eight new nuclear plants in France. None of these investments would ever have survived the commercial light of day. But then they didn't have to.

We crawled through the amazing jumble of pipes and machinery in the belly of the Dungeness machine (incidentally, just recently in 2016, thirty-six years later, given a further ten-year lease of life) – something between a gigantic Swatch watch and the intestines of some giant, metallic robot. Here were locked in billions of pounds and generations of skilled work and planning, the cost of which would never be recovered. But that did not matter to its progenitors. They had built not to cost but to dream. The dream had been cheap and secure power for all, which was going to revolutionize the world's power systems.

Can this kind of super project, this great pyramid-type building, ever be repeated in today's conditions? In today's world, with the computer power and technology available to track and publicize every cost, every expenditure decision, with economies of scale challenged, with decentralization possibilities opening up in every area of economic activity, can this sort of giant undertaking ever go forward? It seems doubtful. Perhaps it can only happen in autocracies.

After that day in 1980, the British Government did indeed have a go, proposing that nine new giant reactors with newer designs and technology than the old AGRs should be built. Margaret Thatcher, the Prime Minister, was enthusiastic. A revived nuclear fleet, generating 30-40% of national power needs, would dish the militant miners.

Only one ever got as far as construction – Sizewell B – and even that one took ten years to be completed.[9] Economics and the mega-sums needed to build such machines made it impossible for them ever to compete, so it seemed, with plentiful gas, cheap and convenient oil, and ultracheap coal.

Hinkley C, if it is ever completed, is probably the last of the line, despite assurances to the Chinese that they can take an even larger stake in the British nuclear capacity in the years ahead. In an age of weaker governments and more intense, exposed,

9. The Cabinet authorized me to make this statement to the House, as Energy Secretary on 8 October 1979. Although the Treasury were hostile, I had the full support of the Prime Minister Margaret Thatcher who saw nuclear power as an escape route from militant coal miners and their rabble rousing leader Arthur Scargill.

transparent, and electronically informed debate over every aspect of public expenditure, a repeat on the vast scale of Hinkley Point C seems highly unlikely. Despite having the extra attraction of producing very low- or nil-carbon electricity – an advantage not even recognized in the 1980s – and despite being able to provide the massive baseload a modern electric world needs, it is hard to see how this Soviet-scale kind of project can ever again get through the thickets of open and fully informed public debate. Nuclear's only chance is to reduce its enormous costs dramatically.

Even if, in the unlikely event, Hinkley Point C is built and completed within budget and at least within a year or two of the yet again postponed 2023 date, it would still be a catastrophe at least for consumers, if not for the investors. As Simon Taylor's fascinating study shows, the financial compensation against risk for the project has been structured in the most expensive and inefficient way.[10] Once completed, the main risk for the investors is substantially diminished. But the consumer goes on paying a high risk premium in the price of the electricity which flows for many years to come. By the time this period ends thirty-five years hence, the cost of electricity from other renewable sources, such as wind and solar power (maybe by then with efficient storage), will have fallen far below the Hinkley price of £89.50 (or £92.50 if another reactor at Sizewell in Suffolk fails to be built) – this figure being indexed and therefore in thirty-five years' time likely to be much, much greater in nominal terms.

Moreover, by then the technology of nuclear power building itself will have evolved beyond recognition, possibly with far cheaper and safer methods and with the spread of small modular

---

10. Simon Taylor, *The Fall and Rise of Nuclear Power in Britain: A History* (UIT Cambridge, 2015).

nuclear reactors built at a fraction of the Hinkley cost and producing electricity at infinitely lower prices.

Hinkley will stand as a monument to flawed beliefs, misunderstood theories, false hopes, and bottomless government naivety and incompetence on an Ozymandian scale.

Despite the very real barrier of sheer cost, to which this all boils down, the World Nuclear Association, perhaps predictably, continues to forecast that between now and 2030 some 266 new reactors will be opened round the world. The cost is estimated at $1.2 trillion. What cannot be answered is the question of where these gargantuan funds can possibly come from – unless the whole cost pattern of nuclear power construction changes radically and massively in a downward direction. This means radically reducing not only the costs of construction but the costs of reducing public fears, the costs of waste handling (possibly by the obvious route of developing very low or non-waste-generating technologies), the costs of robust electricity grid systems, the costs of regulation and inspection, and the costs of eventual decommissioning. None of these things is in prospect or even in the visible innovation pipeline.

One final avenue may yet be open for low-carbon nuclear power – at any rate in democratic and open societies, or even in the consultative or guided democracies round the world. This avenue leads to small modular reactors. By borrowing and developing the technology used in nuclear power units in submarines, the possibly opens out of far less costly, less centralized, and much, much smaller reactors in numerous locations nearer towns and industrial operations.

The most obvious hurdle – public fears about safety – would remain, but the problems about long construction time, immense funding needs, very long-term public commitment, and general visibility would evaporate. The need to construct or renew nationwide infrastructures to accommodate the 'big beast' of massive, centrally generated power would vanish. The inflexibility of central power stations would be replaced by the flexibility of dispersed and local electricity sources at both industrial and domestic levels.

An SMR reactor could cost less than £1 billion and produce about 300 megawatts of electricity. Compare this with the plans for Hinkley C, which would produce about five times as much, but twenty times the cost at least. The technology required would not be all that different from that deployed in building nuclear submarine power units.[11] This explains why Rolls-Royce is interested in further research. American designers are also working with the British at Sheffield University to move to the next small reactor stage. Poland, struggling with the need to reduce coal-fired electricity (and Russian gas dependence), has also declared a strong interest in small reactors, proposing close collaboration with the UK. Since Polish support on another front – EU renegotiation – is increasingly important to the UK, this is plainly one area where energy cooperation can assist broader interests and purposes.

This appears to be the only way that nuclear power costs can come down. The same test applies to all other renewables except that they hold out a stronger promise of getting their costs down as improvements are incorporated. It has become a race against time. Pressure grows daily on governments to reduce subsidies and charges on consumers – pressure that has intensified as fossil fuels have grown cheaper and more reliable and they benefit from new technology that delivers lower carbon per unit of power produced.

11. But the US Navy is reported to be distinctly less interested, not wanting to be drawn into the civil nuclear debate.

CHAPTER FOURTEEN

The China Syndrome

Another Chinese visit – all energy roads lead to China – China, the dominant energy 'player' – China, the biggest coal producer and coal importer – Chinese contradictions – Chinese industrialization path different from the West's – China's worldwide reach in Africa, in Asia, in Europe – emphasis of One Belt, One Road on oil and gas development – green target rhetoric versus hydrocarbon plans and realities.

We are stuck in Hyde Park. It is 1980 and the Chinese are on a large-scale visit to Britain to study our energy policy and our spectacular oil developments offshore in the North Sea.

So lavish were the reception plans for this occasion – a 'first' on this scale from China's Communist leaders – that the whole of Hyde Park is closed to enable the Chinese President and Chairman of the Praesidium, Hua Guofeng, to drive in his official car through the Park from Kensington Palace to Park

Lane. I am accompanying him with an interpreter. The gates at the Kensington end are open, but unfortunately the exit gates on the Park Lane side are locked. Either the park keeper is late or no one has told him. We are stationary, as are all the cars behind in the cavalcade, plus the police escorts. The Chinese President pats my knee and says that when oil gushes out of the offshore installations which are going to be built in the South China Seas, I will be there, specially invited. We exchange further pleasantries, and after a while, the keys are found, the gates, unlocked, and the procession moves on.

In the event, no invitation ever came, nor was there much Chinese offshore oil. Nor did Hua Guofeng remain in office long enough to send it – he was turfed out shortly after his return. He did, however, present me with a rather nice plate. Chinese visits are a bit like that.

Instead of the hoped-for, offshore Chinese gushers, history took a different turn, one which perhaps the Chinese leadership, with their proverbial long-term view, were trying to avoid. Twenty-three years later in October 2013, an historic milestone was passed. The figures showed something which ten years or even five years before would have seemed far beyond credulity. It was the month in which China became the world's largest oil importer, overtaking the United States.

It used to be said that when it comes to energy issues all roads lead to or through the Middle East. In much the same way, it could now be said that when it comes to shaping future energy markets as well as fighting climate change, China is now, and will continue to be, the decisive influence, however fast or slow its growth.

Whether the issue is China's titanic and ever-growing energy consumption, or the development of its own energy resources, or the attitude it takes to its own carbon emissions (now the largest in the world at 27% of the global total), or its worldwide, large-scale involvement in primary energy projects and production on every continent, or its immense shipping capacity (again the world's largest), or its financial system, or indeed its political structure, the conclusion is the same. The policies which China follows will determine the world energy future.

On every energy front, China is becoming the dominant player. It is the world's largest energy consumer. It is the largest coal producer, whilst also the biggest importer and user of coal. It is the biggest importer of oil, having overtaken the United States which now heads towards self-sufficiency in both oil and gas. Yet it is also the world's fourth largest oil producer, after Saudi Arabia, Russia, and the US. It is the biggest builder of nuclear-generating capacity, with twenty-eight new nuclear plants under construction on top of the existing forty. It is the biggest constructor of wind farms. It also claims to have as much shale gas as the US (see more below), and while it has less than half the United States' vehicles (roughly 120 million against 250 million in the US), by 2030 the estimate, on present growth trends, is that it will have 350 million, far above the American level.

To accommodate this flood of cars, it has built in the last two decades, from virtually nothing, 100,000 kilometres of trunkline, or intercity highways, as against the 75,000 kilometres total of interstate roads in the US.

China's stances and attitudes on energy issues are rife with paradox. It is deeply concerned about immediate pollution from sulphur and other smog-inducing emissions, about water shortages, and about longer-term climate changes. Yet it is planning more new coal-fired power stations than the rest of the world put together. (144 is the latest figure.) These are certainly efficient and will be using higher-quality coal, but they are not equipped with carbon capture and storage facilities, or even the necessary modifications to allow future CCS add-ons.

China longs for self-sufficiency in energy, regarding reliance on global hydrocarbon supplies as a dangerous weakness. Yet around the world it has vacuumed up extensive oil, gas, and coal concessions, as though it is going to be importing massive amounts of oil and gas forever. It declares that it has no interest in the labyrinthine complexities and internal instabilities of the Middle East, yet its appetite makes it increasingly reliant on Middle Eastern oil supplies, just as the West is taking less and less from the region. It regards both the US and, so it believes, the US' client state, Japan, with hostility and mistrust, yet it is a huge importer of Japanese goods and home to immense Japanese investment, while its promised shale gas development relies extensively on evolving US shale technology.

The contradictions continue. China shies away from reliance on foreigners and draws in on itself politically as the Middle Kingdom, while seeking full integration with the global trading system through its membership of the World Trade Organization. It eschews and denies any imperial or expansionist ambitions while inching its territorial sway outwards into highly contentious and strongly disputed areas in the East and South China Seas. It insists on its complete commitment to peaceful development while acting often in a highly provocative and even aggressive way, as is the case with its suddenly renewed claims over the Senkaku Islands (Diaoyu Islands), and its insensitively timed designation of a large surrounding area as an Air Defense Identification Zone (ADIZ). It takes a rigid and absolutist view of sovereignty over Tibet and Taiwan, yet it has allowed Hong Kong amazing latitude, including its own currency and even its own involvement in aspects of global affairs – namely through the Commonwealth, the World Health Organization, and other bodies. (The 'Umbrella Protest' of autumn 2014, which brought tens of thousands of Hong Kong students onto the main Hong Kong island thoroughfares, was the point where this latitude collided with other global currents, as well as with a kind of power cut in comprehension by British Members of Parliament, who seemed to have forgotten that Hong Kong was part of China.)

And so the list goes on. In short, as we peer into the energy future, it becomes harder and harder to assess which way China will go, assuming it remains one centrally directed entity capable of reaching and enacting consensually agreed policies and strategies.

The Chinese Way

Chinese carbon emissions per head are low. Of course, in total they are enormous – the largest of any country in the world. But per head, they are about five times lower than US levels and about a third of the average EU level per head.

The common assumption is that this will change, that China will follow the Western industrial path, and emissions will rise rapidly, just as oil experts assume Chinese oil consumption will rise with further industrialization. Oil consumption per head is *nine times* higher in the US than in China. That is what has happened in the past in the West, so the inevitable (and facile) forecast is that the Western past will be extrapolated into the Chinese future.

There are at least two reasons why this common assumption will be proved wrong. The first is that China will leapfrog whole stages of Western industrialization. Advanced technology will take the economy and the energy use pattern straight into the mould of the future with vastly improved energy efficiency and reduced energy intensity. CO_2 emissions will stop growing so fast even though tens of millions more Chinese will join the richer consumer classes annually, owning tens of millions more vehicles, because China will cut out the gas guzzler motor age. The huge leap in demand for more petroleum from the transport sector will just not occur.

In other words, China is simply not going to pass through what Naomi Klein labels in her book, *This Changes Everything: Capitalism vs. the Climate*, its 'Dickensian phase'. (Oddly Klein then goes on to propose steps for China which would send it back precisely in that direction, but, as they say, 'chacun a son goût'.) Her bizarre argument is that since existing climate-combatting policies are plainly not working (CO_2 emissions having risen, she observes, by 61% since 1990), the answer is to block international free trade. This she equates with something she calls 'market fundamentalism' – a bugbear which does not, in reality, actually exist.

Her idea is therefore to keep China poorer, ensuring that half a billion Chinese continue to live in mediaeval poverty, burning in their homes and choking on carbon-loaded biomass fuels, and that China carries on with its trajectory of increasing coal reliance. China would thus be stymied from bringing a further 200-300 million people into a more middle-income lifestyle, where energy use efficiency could be maximized, local and low-carbon energy processes, more readily adopted, and modern, much less energy-intensive economic activity, expanded to meet more sophisticated consumer choices.

This seems to be getting things the wrong way round. China has already engineered the greatest fall in world poverty in all history. Surely it should not be stopped from doing more.

Meanwhile, as China grows wealthier, the country's natural gas demand, starting from a low level at about 5% of total energy consumption, will push against both coal and oil, thereby delivering a steadily lower-carbon dividend. The declared Chinese official aim is to raise gas from 5% to 10% of primary energy use by 2020.

None of this will have much to do with the European decarbonization 'example', of which the Chinese are probably hardly aware, for which they care even less, and which is anyway statistically insignificant in reducing global emissions. If any Western example is making a mark, it is that of the American shale gas revolution and the enormous carbon-cutting switch from coal to gas that has occurred there. China wants to emulate the American shale miracle, in theory it has the shale gas (and maybe the shale oil) resources to do so, and is happy to call in American shale expertise to kick off its own shale boom, copying its technologies no doubt later. In practice, Chinese shale development will be much slower and the gas expansion will also have to be fed by LNG imports (a lot from Australia) and by pipeline contracts from Central Asia. But there will undoubtedly be a larger energy share from gas, both for transport and for industry (within a growing total of all energy sources), and this will help slow carbon growth, as well as ordinary atmospheric pollution, in due course.

The second reason why China may perform better than past Western societies is that the Chinese model will be different. The declared guiding purpose will be to avoid unrestrained and debt-financed hyperconsumerism. The slogan of the Fourth Party Plenum in November 2014 was 'Enough'. For Chinese people, declared Chairman Xi Jinping, three pairs of shoes will be sufficient.

This whole approach is being encapsulated in another one of the phrases or slogans of which the Chinese leadership are so fond – 'The New Normal'. This is meant to be the embodiment of a spirit of restraint and prudence in the pattern of economic advance, contrasted with Western excess. With this is intended a rising trend of energy efficiency, a slower trend of energy demand growth, and the emergence of a greener pattern of energy production. This is laid out in a slightly portentous-sounding document – *Energy Development Strategy Action Plan (2014-2020).*

In essence, and somewhat in contrast to the reality of today, this is saying that the previous (and current) extensive international search for energy sources will be replaced by a new focus on domestic energy production. Much less clear is how this is to be achieved, and whether it is in any way consistent with China's worldwide, and spreading, involvement in the infrastructure of the entire planet.

The Chinese undertaking, made in a 2014 fanfare agreement with the US under President Obama, is to have carbon emissions peak in 2030. This will probably happen anyway as a direct result of the policies and technology trends described above. What the Chinese are agreeing to is what will happen, assuming their nuclear programme stays on track (as already noted, with eight new reactors in 2015 and a further twenty-six under construction). If that falters, then the enormous fossil fuel consumption will climb on past 2030, and so will emissions. 2030 will become 2040.

In this picture being painted of a calmer and cleaner energy future, there is bound to be an element of self-delusion – both in Chinese thinking and in the Western idea that example and self-sacrifice elsewhere will somehow help bring about the promised transformation. Reality points quite the opposite way. In harsh reality, new coal plant capacity added in 2013 alone exceeded new solar capacity by seventeen times, new wind capacity by four times, and even new hydrocapacity by three times. This is an economy that is going to remain predominantly coal-driven for decades to come, however much the low-carbon alternatives are developed and however eloquent the assertions of party leaders in Beijing.

Recent evidence suggests that the use of coal in China may be much higher still than anything so far assumed or disclosed. A Norway-based research centre has suggested that in very recent years Chinese coal-burning may have been 17% higher than previously announced. In terms of CO_2 emitted, that would amount to some 900 million extra tonnes of CO_2 – more than the total annual emissions of either the UK or Germany.

Observers of the Chinese scene seem constantly surprised by the sheer size of it all, and the way in which the Chinese numbers, whether of energy production, energy imports, energy consumption, air pollution, or carbon emissions, tower over the statistics for all other nations (except, in some respects, the US), making them look puny and insignificant within the global totals.

The mistake, and the roots of many misunderstandings about China, is to treat it as just another nation on the economic and diplomatic chessboard. As Martin Jacques explains, 'China should not primarily be seen as a nation-state but rather as a civilization-state. The implications of this are far-reaching; it is simply not possible to regard China as like, or the equivalent of, any other state'.[1]

This applies as much in the energy sphere as everywhere else. The volumes and the impact on the whole global energy network are simply of a different order to those of most other countries.

1. Martin Jacques, *When China Rules the World: The End of the Western World and the Birth of a New Global Order* (Allen Lane, 2009).

In a moment we will look at how Chinese energy-related activities are spreading far and wide across the whole planet with fundamental geopolitical implications. But it is worth noting in passing that this phenomenal dominance seems almost too big for some Western policymakers and thinkers to grasp. For example, it is a bizarre fact that Britain, whose total pattern of energy activity, influential though it once was, is now super-dwarfed by the Chinese, is actually convinced that it can help and push the leviathan along its way.

Incredibly, widely unacknowledged and substantial amounts of British consumers' and taxpayers' resources flow directly into Chinese company coffers through subsidies to wind power and nuclear developments in the UK. These are both areas where Chinese investors have taken big positions and must be greatly enjoying their returns, courtesy of British energy policy and those who fund it. China General Nuclear Corporation has purchased an 80% stake in wind farms near York, Newcastle, and Peterborough, and is eyeing offshore stakes as well. And if these outcomes for China's benefit are in a sense indirect, in that the fat pickings are available to all outside investors (Norway being another good example, through Statoil with its very large UK offshore wind farm holdings), other more marked and conscious support flows direct from the British Government.

It requires no genius to see that while this works in theory, it cannot possibly work in practice. The sums are minuscule when set beside the Chinese colossus. The momentum of Chinese energy production is far too great to be checked even momentarily by this pocket-sized (for the Chinese) kind of gesture. China's own highest aim, which may not be achieved, is to reduce coal use as a proportion of total energy demand from 68% to 60% by 2030. The latter figure is of course a percentage of a much larger total energy consumption by then, and therefore a far larger amount in absolute terms.

Even to do this, it will need to build an *additional* 800-1,000 gigawatts of nuclear, wind, and solar electricity capacity, as well as switch even faster away from coal to gas. That is close to the whole generating capacity of the US today and around ten times the total generating capacity of the UK (and fifteen times its current output).

The great hope of the Chinese is that much of this extra gas appetite can be satisfied by home-produced shale gas. In fact, even if all shale gas targets are reached, China will still need to be importing very large volumes of gas via pipeline and LNG tankers. Its total gas demand will jump from around 100 billion cubic metres a year today to 635 billion cubic metres by 2035.

In practice, the shale contribution could be far less. Chinese shale resources appear to lie much deeper than American deposits, and in more difficult terrain to boot. In Brussels I met up with Liu Tiafen, the Chinese deputy energy minister and his team of fourteen aides. Mr. Liu was there at the Berlaymont to outline China's five-year plan for shale gas development. China, he said, had even bigger shale gas resources than the United States, some 68% more in theoretically 'recoverable' quantities than the Americans. There were, he conceded, 'some geological problems' in accessing and recovering some of it, but there was now in place a five-year plan to expand shale gas production and thus contribute to China's energy strength.

Afterwards, one of his aides quietly explained that these 'geological problems' were not just that much of the drilling would be needed in remote and very mountainous areas (in contrast to the accessible US sites); there were other problems as well. If it was not mountains, it would be people. Some of the big deposits were located right under Chinese cities and very crowded areas, so that finding space to start hydraulic fracturing operations would be an added difficulty. But China, he assured me, would overcome all these obstacles.

Of course they won't. Quite aside from difficult and deep sites, the whole vast infrastructure of machinery and equipment needed, of ample water supplies, of pipelines to get the gas away, and of distribution, is just not there and will take years to develop. The criss-cross pattern of distribution pipes across the US is not there either. Nor is the army of small- and medium-sized shale gas enterprises that have given the American scene its boom qualities. China will be relying much more on big state companies, riddled with slow-moving bureaucracy, although in the latest licensing rounds, Chinese private firms have been allowed to bid for blocks and minority foreign investors have been allowed in.

In the case of water, already in desperately short supply and needed in gargantuan volumes to do successful fracking, the problem may be insoluble.

The Tentacles Reach Out

It is therefore on imported pipeline gas and seaborne LNG that China will be relying, and this is where Chinese needs and outward investment flows mesh directly with the wider global energy scene.

China is already connected up via pipeline with Turkmenistan and Kazakhstan. Its national companies are busy building pipeline connections with Myanmar and a number of Central Asian states. A multibillion dollar deal to supply gas for thirty years ahead has been signed with Russia, although both the price and the technical details remain obscure. It comes after years of negotiation between China and Russia and the suspicion must be that in the end, Russia acceded to Chinese pressure for lower prices. It needed the deal badly for presentational purposes in May 2014 to demonstrate to EU countries discussing sanctions that it could sell its gas elsewhere. Or that was the idea. Whether it works in practice, and whether the gas itself can be successfully produced – it will come from new fields mainly in Yakutsk and difficult areas adjacent to the Arctic – remain more questionable.

But both import pipelines and expanded sea routes for LNG will be Chinese tentacles as well as Chinese feeders. Everywhere across Central Asia and into Afghanistan and down into the Indian subcontinent, Chinese groups are building, buying into, and planning local infrastructure, local support services and facilities (sports stadia are a favourite), local rail terminals and railways, local ports and shipping. Across Africa the pattern is the same, and again in Latin America.

Half of African exports to China are now oil and account for 30% of China's huge oil import total. China has developed oil interests in Algeria, Angola, Sudan (very extensive ones), Congo, Nigeria, and Mozambique. It also has built up big mineral interests in Zimbabwe and Zambia. The local effect has not always been harmonious. Back in 2007, the Zambian presidential candidate Michael Sata complained that Zambia was becoming 'a province – no, a district, of China'. Later, when he stood again successfully, he became more emollient and friendly. Money talks.

There are now very few countries where the Chinese presence is not growing fast. China is now the best customer for oil from the Gulf States. China has big projects under way in Saudi Arabia. China is Iran's big customer. China is the biggest outside oil investor in Iraq – bigger than either BP or Exxon or Total.

Brazil is also set to be a major oil supplier to China. Sinopec and the China Development Bank have a deal with the Brazilian national oil company, Petrobras, to lend it $10 billion in return for 200,000 barrels a day of crude for ten years (a company now in deep domestic and political trouble). Earlier, China had bought a 40% stake in Repsol Brasil, one of Brazil's largest private oil companies. Chinese oil links have also been built up with Ecuador, Peru, Chile, and Argentina (said to have some of the largest shale gas reserves after the USA), Mexico, and China itself.

Europe has not been left out of the new growing mosaic of Chinese interests. Chinese investment and involvement in the UK has already been described, and heavy signals are being sent suggesting they want to invest much more – hardly surprising when one considers what a sweet and reliable deal they have already secured with EDF at Hinkley Point. There must be more, they are surely saying, in this honeypot.

Chinese officials have also been in Warsaw showing interest in Poland's shale gas prospects. Prime Minister Li Keqiang was there in February 2014 talking of China taking a large stake in Polish shale development, and indeed in shale generally throughout Europe. Later in the year, he was in Belgrade discussing European infrastructure and outlining plans to turn the Greek port of Piraeus into a hub for Chinese trade with Europe. Chinese plans to link itself with Central and Eastern European markets include upgrading Greece's dilapidated railway system and funding and building a new high-speed bullet train link, starting in Greece and passing 'through Macedonia, Serbia to Budapest and further towards Western Europe', according to Hungary's current Prime Minister, Viktor Orban.

Add in already commenced projects to build new power plants in Serbia and Bosnia, a highway to Croatia's Adriatic coast, and an already completed and opened one-mile Serbian bridge across the Danube, and a picture emerges of Chinese commitment to, in Li Keqiang's words, 'a massive express line through the Balkans to China at lower economic costs'.[2]

All these developments, some merely putative, some under way, are seen as part of the portmanteau idea of the New Silk Road, the highway which will bring Chinese goods to the heart of Europe, passing through Central Asia on the way and bringing the whole vast landmass of territory lying between China itself and the edges of the West into new phases of development and growth.

Sloganized into the phrase, 'One Belt, One Road', the Silk Road thinking is for both new land and sea routes, opening up East Asia and linking China with European and Western markets on a gargantuan scale.

The land proposal is for an economic corridor beginning at Kashgar in Xinjiang Province in China, 3,000 kilometres away, passing through the 'Stans' and pausing for breath, so to speak, at Istanbul, on the edge of Europe. From there Athens' port of Piraeus (in which China already has taken a significant interest) could be the next stop, while plans (already mentioned) to finance and build new railway links from both Bulgaria and Greece up into the heartlands of Balkan and Eastern Europe have also been announced. The scheme involves a quantity of new power plants in Pakistan as well as large-scale investment in a deepwater port at Gwadar. This is already well under way.

The ambition is breathtaking. Not all of it will happen and the whole expansive vision does not quite tie in with the parallel

2. Wang Zhao and Pool, *Reuters,* 17 December 2014.

tableau of a more restrained China looking modestly after its own affairs and eschewing any idea of interference with others. Nor will China necessarily get the full support and approval of other Asian and Pacific powers in carrying forward its magnificent-sounding plans. Neither Australia, nor India, nor the US, nor Japan, has shown much enthusiasm. On the contrary, they have expressed extreme wariness.

This has been exemplified in the coolness which these nations and others have shown towards the proposed financial and funding agency tasked with carrying forward the immense portfolio of infrastructure and energy development involved. This agency, the Asian Infrastructure and Investment Bank, has neatly split advanced world opinion, with Britain leading the way and other European powers agreeing to be founder participants in the new institution. The US, Japan, Australia, and Canada have stood firmly apart. The Japanese, in particular, see European support for the Chinese initiative as amounting almost to betrayal by their supposed friends, the British, and pandering to Chinese political motives for expansion hidden behind the commercial screen of the new bank.

Lawrence Summers has gone further, asserting that the new bank marks the end of more than half a century of American dollar dominance of world development, exercised through the 1945 Bretton Woods institutions. But all these dire warnings and fears depend on the extent to which Chinese plans are in the event fulfilled and how long it all takes. Some will be, especially the big energy projects, but not all.

Perhaps the most relevant point is to note that the Chinese vision rests heavily on the exploitation of the immense commodity resources – oil and gas in particular – to which the new roadways will give access. How this squares with worldwide commitment by governments to phase out oil and gas completely is just one more mind-bending contradiction.

The New Chinese Web

All around the Indian Ocean, China is weaving a new web of commerce, transport, and energy supply. The mixture is one of trade in goods, large shipping fleets to move the merchandise around and across the whole sea region, and security ports and bases to protect both the ships and the goods – and the oil and LNG flows heading China-wards.

This emerging pattern makes a statement. It is that the Indian Ocean, rather than the Atlantic Ocean, is becoming the central pool of global power and progress. China is building ports and bases on the East African coast (at Maputo, at Zanzibar, and on the Somalian coast). It is well advanced with its major port projects at Gwadar in Pakistan and at Yangon. It has a big trading base foothold in Mauritius. The Indian Ocean is metamorphosing, almost without the world noticing, into the Chinese pond.

In Martin Jacques' words, the Chinese appetite for commodities 'has no parallel in global economic history'.[3] The way China goes, and the way its energy flows and systems develop, will have massive impact on all our lives, and not just in decades ahead but very soon – in fact, now. To summarize:

> It is Chinese growth that will either save the world from, or send it into, the next global recession. Its ups and downs will phase us all.

3. Martin Jacques, *When China Rules the World: The End of the Western World and the Birth of a New Global Order,* pg. 597.

It is Chinese foreign policy which will determine whether rancour and enmity with its neighbours will make the Asia-Pacific region, now accounting for a quarter of global GDP, a zone of continuing prosperity or conflict and slowdown. (This will be shown among other things by how far China pushes the Senkaku Islands quarrel and other island-annexing or island-creating initiatives in the South China Sea, and how China reacts to the increasingly Hitlerian scene of tyranny and bloodshed in Pyongyang.)

It is Chinese oil and gas consumption, and its commodities hunger, and use, which will determine future hydrocarbon prices and the prices of others.

It is Chinese policy towards its currency (the under-priced RMB) which will balance, or dangerously unbalance, world trade and financial flows.

It is Chinese, not European, policies and practices which will decide whether humankind can make an effective contribution to climate change control and world decarbonization.

It is Chinese capital investment and technology which will influence the pace of poverty eradication in most continents and even underpin Western investment requirements.

It is Chinese diplomacy and external attitudes which will help or hinder the damping down of the smouldering and splintering Middle East.

It is China which will decide whether the rebalancing between non-Western and Western power and influence in the world is to be a peaceful process with benefits all round, or an ugly escalator of global friction and destabilization.

The argument is put forward, especially in Washington circles, that all this must somehow be stopped or 'managed'. China's ballooning trade with the rest of the world must be interrupted and curbed. China, the greenest policymakers demand, must be halted in mid-modernization since it is the chief source of rising emissions.

But this is nonsense. China being held back is the surest way to continue massive emissions, energy inefficiencies, and pollution-spewing industrial and agricultural processes. China leaping forward, with rising incomes per head, is the best and only way in the opposite direction – towards new and greener technologies and innovations, and the lower-carbon future its leaders have assured the world they are aiming for. In the end, the giant 'civilization-state', in its own undoubted interest, must sit down and find the most harmonious place it can with all the networks of 'nation-states' to which it is inextricably connected.

In the end, the planet has to accommodate the giant, learn from it, live with it. In the digital age of hyperconnectivity, there will be much to learn from China as to how a stable system of government evolves on such an enormous scale, how it reconciles harmony and stability with prosperity and economic expansion, how the need for enterprise and initiative can be combined with central control, and how it keeps the motor of innovation spinning – without which no progress will occur.

To grasp and understand what China's energy policies and strategies mean for the rest of us (and we need to understand to prepare ourselves and adjust our own energy decisions), it is necessary to peer much deeper into the Chinese attitude and state of mind – which is itself the product of millennia of experiences and changing fortunes, culminating in the century of humiliation and ignominy which brought an impoverished and invaded China to its knees. After that came the hope and agony of Mao's colossal misgovernment. Then followed, like a fabulous garden suddenly blooming out of rubble and dirt, the amazing modernization process which has lifted China, in a couple of decades, from humiliating poverty and stagnation to the topmost ranks of modernity and industrial power.

A certain contradiction lies at the heart of Chinese attitudes, difficult for the Western mind to grasp. Even while they encourage planetwide investment and the acquisition of commodity supply chains, the rulers of China regard high dependency on imported items – especially oil – as a weakness and vulnerability. This translates into an ambiguity – a desire to secure resources but a reluctance to be sucked into the politics of the regions concerned. The commercial footprint will grow in the Middle East, but China will resist deeper security or political engagement. It may find this balance increasingly hard to maintain.

Our fate is now largely in Chinese hands. So is the future of the entire global energy system. So is the struggle to combat dangerous and destructive weather patterns which could paralyse the planet. The wisest goal for the international community is to let China move naturally to a greener, safer world – and take the rest of us with it.

CHAPTER FIFTEEN

The Tomorrow We Deserve

The pathway to a better energy future – the pathway blocked – the recognition of error, a first step – zealotry, the enemy of progress – technology, the real driver of change – harnessing green capitalism – the unavoidable world expansion of fossil fuels – the absolute imperative of cost-effective carbon capture – cooperation between green and hydrocarbons worlds essential – turning energy abundance into cheap, reliable, low-carbon energy – the next ten years – the unending energy revolution – will it ever be dull?

What is to be done? Somewhere ahead lies a balanced and harmonious energy world in which cheap and plentiful power is accessible to billions of people, in which electricity flows reliably, in which carbon emissions into the atmosphere and poisonous air-polluting substances are minimized, in which maybe even global warming is curbed within the 2°C above 1990 levels, which scientists insist is the catastrophe threshold.

In the course of this book, I have argued that the road to this prospect is strewn with obstacles, pitted with ill-designed and counterproductive policies, blocked by impatient zealotry and blind dogma, and scattered with misunderstandings and missed chances. The cost barriers are horrendous. The route may even have become impassable.

Britain and Germany provide the most vivid examples of muddle and misdirection, taking us away from, instead of towards, a better energy future. But European energy policy as a whole sets the wrong context for reform, while for millions in the poorer regions of the world, the prospect of secure, cheap, and clean energy is receding.

As Chapter Ten showed, there is no one leap or escape from the labyrinth, no sharp correcting turn to wheel round to drag us onto the right road. Nor are the really decisive moves in the hands of any one government or society. But the conclusion of past chapters is that some sensible things can now be done to alter course and bend trends and events in a better direction.

We have tried to explain that the journey to a better pathway begins with the admission of error. In a way, this should be the easiest step because the evidence of failure is so striking and because both camps, green and hydrocarbon, agree that present policies and trends are not working. When the crusaders for 'climate justice' hailed the 2015 Paris Agreement as 'a new beginning', that is of course just what they were saying. Nothing, or very little, so far has worked. This has not been enough to deter some voices from claiming that the past energy decade has been an outstanding success despite all facts to the contrary.

The second step on this journey is to see that zealotry is the enemy of progress. It is the rush to war with all fossil fuels, and with the existing energy structure of the planet, almost regardless of the social and economic disruption, which has set the whole transforming process back.

The third step is to accept that new technologies and market forces are by far the best drivers of change – change in the production and the distribution of energy, and crucially in the efficiency with which energy is consumed. A fourth step is to see that finding cleaner ways of burning fossil fuels will prove far more constructive than trying to stamp them out. This is the assignment between green ambitions and the reality of what is actually happening which must be made.

Fifthly, there has to be cooperation between the two energy worlds. This has been a central theme of *Energy Empires in Collision* and the lack of it remains the greatest danger.

As to energy shortage and energy scarcity, there is none. Energy sources of all kinds, from the earth, from the sky, from the sea, from the land, are in infinite abundance. Fossil fuel prices are low and set to go lower for a long while ahead with major global political repercussions. Ways of harnessing, delivering, and using energy and power to meet the needs of all, while respecting the fragility of the planet, and if we are clever, while actually countering threats to the planet, are within the reach of technology, of research, and of both governments and the leading world energy companies. This would be real progress, not posturing. This is a story that has no end.

1. Admission of error

'Our current path is getting us nowhere', writes Dieter Helm, the master craftsman of energy policy, who has shown with such devastating clarity how much is going wrong and how it could be put right.[1]

Page after page of Helm's work catalogues the errors, forced and unforced, the misplaced hopes, the false assumptions, the twisted arguments, the unintended consequences that have come to bedevil both the European and the world energy scene, and have already led to so much unnecessary hurt and danger across most of the globe.

Helm unmasks the bald facts. Carbon, he observes, 'is being added to the atmosphere at an accelerating rate.' In other blunt words, all the climate policies, the agonies, the price cruelties, the job-losing costs, the landscape despoliation, the frightening uncertainties are failing. The predictions, he adds, about the oil running out and about ever-rising fossil fuel prices were just wrong.

Like other commentators, however expert, Helm is inclined to underestimate the sheer scale of the difficulties which democratic governments and policymakers face in trying to escape myth and confront reality, or the maddeningly long time it takes to get things changed in the great energy world complex, especially when a thousand day-to-day pressures and distractions are in the way.

There come to mind such seemingly out-of-the-blue events as the juddering world oil shocks of the 1970s and 1980s, the attempt by aggressive coal miners led by Arthur Scargill (a truly malign influence) to halt British power stations, nuclear power station accidents such as Chernobyl, the more recent near-nuclear disaster at Fukushima, the sudden eruption of nihilistic terror across large parts of Iraq and Syria, the massive oil leak

1. See the concluding chapter of his book, *The Carbon Crunch: How We're Getting Climate Change Wrong – and How to Fix it* (Yale University Press, 2012).

and fire at the Macondo site in the Gulf of Mexico, or Libyan anarchy and terror in the Algerian desert. The list stretches on grimly.

All these kind of events, and many more, add to the uncertainties which catch even the wisest and most far-sighted administrators and most percipient forecasters totally unawares. Indeed, perhaps the greatest failing in the world of energy and climate experts has been to be too certain, to assume too many 'given' trends, to put too much weight on supposedly scientific predictions which are in fact loaded with simulations and plain guesswork.

Something dangerous is plainly happening to the world's climate and there may be not much time left to check it. This was the one clear and agreed message to emerge from Paris. But can the precise date for the rendezvous with disaster, 2050, really be so precise? A well-known hymn reminds us that for God 'one thousand years is but a single day'. Supposing the scientists are a day out. Who knows?

All who presume to shed a little light on future events are inevitably vulnerable. But it is a question of tone and degree. The future must always carry a health warning. Those who omit it are courting contempt, and deservedly so.

In particular, scorn must be reserved for the numerous forecasts at the beginning of 2014 – forecasts from bank analysts, energy company economists, academic consultants, columnists, the whole galère with a few brilliant exceptions. Yes, even from the IMF and even from the official OPEC forecast, *World Oil Outlook* – forecasts that crude oil prices would stick at around $100 for the coming year, indeed for the rest of the current decade. Wrong, quite wrong – and probably extremely expensive for many firms and many investors.

Its Not That Simple

But on Helm's side are the facts – many of them identified and expanded upon in the previous pages of this book. It is simply not possible to refute the fact that carbon emissions are continuing to rise, both in Europe and worldwide. It is not possible to ignore the fact that coal-burning for electric power is massively increasing. It is only complete and willful blindness that can hide the facts about rising energy poverty and cripplingly uncompetitive energy costs for industry. It requires a state of total denial to fail to see the dangers of weak investment in new power plants, and the likelihood of severe shortages of power ahead. It demands a complete departure from reality to see oil, gas, and coal being phased out by 2050 and renewable energy sources providing the massive flows of electric power on which the future world will depend. It demands fairytale belief to assert that such a vast undertaking as the transformation and decarbonization of the entire world can be achieved inexpensively and with only minor sacrifice. On the contrary, as earlier chapters have illustrated, the colossal investments required to bring about the transition, and the cutbacks in current consumption to make room for them, will be very painful indeed.

Perhaps the very word 'transformation' is the wrong one. It implies total victory for renewable energy sources and total defeat of all fossil fuel activity. But the central argument of this book has been that the two empires should work together rather than fight, even as both transform. Warfare between them, each trying to undercut the other, will result in zero gains for either side and defeat for climate goals overall. That would be the greatest betrayal. As in many forms of modern warfare, it would be the millions of civilians who suffer, while neither side achieves victory.

Helm says it is time to square with the voters. It is also time to square with the governors, with the policymaking establishments, with investors of people's savings and pensions, with the poorest families and the weakest in our societies, and with the millions who have no access to power or to clean water.

Energy and climate issues are too complex, too delicate, too interwoven to withstand clumsy and partial policy interventions with totally unforeseen and usually highly damaging and counterproductive consequences.

2. The Rush of Intolerance

So let us put aside the falsely reassuring fairytales and start with blunt realities. In the global battle to curb greenhouse gases and combat climate change, the contribution made by the European nations makes an insignificant difference. The struggle to limit carbon emissions passing into the atmosphere will be won or lost, as it is being at present, in India, in China, in Russia, and also in the US – where emissions are falling, and ironically far more progress is being made than elsewhere in the West, but emissions are still by far the highest per head worldwide.

The awkward truth which is not being told is that the harsh and self-harming measures being supported by Western and especially European authorities to impose very high energy prices on homes and industries, the calls for 'accelerated' penalties on fossil fuels, thus switching large swathes of income from hard-pressed families and struggling workplaces to big investing corporations and wealthy landowners, even if they were effective – which they are not – add only insignificantly to the struggle against climate change. They may even make matters worse. The numbers are minuscule.

The contribution which countries such as Britain or Germany *ought* to be making to address climate change is quite different from the one they are making now.

The billions being poured into subsidies for uneconomic renewable energy – and, in particular, into vastly expensive wind farms – should be concentrated on developing the new technologies to get the costs of greener forms of power production down and to deliver the massive flows of cheap electricity which these nations are absolutely determined to have. That is why the Global Apollo Programme, already mentioned in the renewables chapter, should be strongly supported.

Instead, all the official policymakers and their critics, itching to lay out one set of measures or another in both the energy and climate change fields, should stand right back, encourage the two 'empires' of fossil fuels and low-carbon to cooperate rather than attack each other, concentrate on the broad tasks of maintaining stability, keeping the peace, upholding the law, encouraging enterprise, and regulating markets, and leave the whole evolving energy process to be shaped by the onward march of technology, by innovation, by unending human ingenuity, and by the forces of markets and consumer need, desire, and choice.

Taxes on all profitable enterprises there have to be – to fund the government's essential duties and services. These will always have to apply to companies in the energy business as much as to any other sector. But once taxes turn discriminatory and are based on highly subjective judgements about supposed 'externalities', the trouble begins. Taxing becomes a sort of targeted moral crusade. The truth is that all business activities have external impacts. All impose costs and requirements on society and the community which governments identify and seek to remedy, and for which they need funds to do so. Very nearly all economic activity, both production and consumption, generates carbon dioxide. All should be taxed as evenly as is possible and reasonable.

Because this is the mixture which is going to deliver what the policymakers, as they fiddle and probe and prod and adjust the energy system, will never, ever succeed in achieving. There are just too many possibilities, too many what-ifs, too much entanglement between the energy supply and demand chain, and the politics and destinies of too many societies, tribes, states, global interests, for human minds to encompass it all.

3. The Real Drivers of Change

Instead, there has to be something for which there is a real demand, a real want for which people are prepared to pay. Politicians and experts cannot create what is not there, can no longer make people want what they do not want and are not prepared to pay directly for, even if in the past, long before the information age, a degree of mass brainwashing was possible. If the demand is there, the supply will be there to compete for a share of it. Agile brains and skilled minds will devise ways and means, products and processes to satisfy that demand at a competitive price.

This is an age in which the pace of technological advance has never been faster, the possibilities of ingenious new advances, never greater – in the energy production and consumption fields as much as elsewhere. The longing for progress to be sustainable, for the world to cope well and in a kindly fashion with its prospective nine billion (or more) human souls, is universal. An age of cleaner, greener, and cheaper energy is desired and can be delivered, but not by a ragbag of energy and climate interventions, applied in a crazy patchwork of sticking plasters which overlap and peel off. By far the best delivery vehicle will be the low and competitive cost of green energy sources and equipment.

A message the story of *Energy Empires in Collision* has sought to convey is that green energy and market capitalism can be, or should be, united in the same cause, making it profitable, economical, and desirable both to handle coal, oil, and gas production and consumption in low-carbon-emitting ways AND to develop alternative low-carbon sources and new green technologies which are affordable and competitive – or at least to find a balance between the two worlds. And of course the strongest driver towards increased and cleverer energy conservation and efficiency can be the greatest 'alternative' energy source of all.

4. Clean fossil fuels

Whatever happens, whatever the 'commitments' by governments, whatever the passion, the oratory, the marches, the indignation, the warnings, billions of tonnes more coal are going to be burnt, hundreds of millions more barrels of oil will be consumed, trillions, even quadrillions more cubic feet of gas will continue to be used worldwide. No taxes on carbon, no levies, no cap and trade regime, no accelerated closures of coal plants by order across Europe, no end to deep coal-mining in Britain, no bans on coal use in Germany, no warnings to the Polish Government, no wagging of fingers at Russia, nor even the bankrupting of giant US coal producers – none of this is going to stop or even phase out fossil fuel burning, whether by 2050, 2070, or 2100. These are the bald, visible, undeniable, inevitable facts of what is happening, as page after page of *Energy Empires in Collison* has explained and documented.

Unless a means can both be discovered, engineered, and successfully introduced into the global energy system for extracting and somehow disposing of the resulting CO_2

emissions, at least 1 trillion tonnes more carbon will go into the atmosphere by 2050, notwithstanding the continuing and healthy carbon reabsorption in the seas and forests of the earth.

There can only be one conclusion. Either a means has to be found – a practical means and a means consistent with the imperative of affordable power for the poorest nations and societies on earth – of limiting the carbon emitted from what is necessarily expended, or the catastrophe predictions of science will be put to the test. This is the inescapable rendezvous with reality that must lie ahead.

We have seen how there are various carbon-limiting possibilities, some of them less expensive, such as the simple idea of sharply increased efficiency in coal power generation, some of them still far too expensive for government funding to bear or for developing nations to afford. Possibly, ingenuity and technology will turn round a corner and deliver cheaper CCS methods. Possibly, carbon can be sequestered without the added cost of siphoning it away over long distances (to empty or near-empty oil fields offshore). Possibly, the added cost of CCS can be spread out into future generations. All have been suggested; none has been yet applied. The story continues.

5. Green Capitalism

The green instinct has all along – so far – been towards more intervention, more taxes, more charges, and more regulations placed on fossil fuels at every stage in the chain of production and use, combined with more support for low-carbon alternatives. Cost is the lesser concern. The mantra has been that the price signals of market capitalism must be defied and defeated, and the consumer, brought to heel. Market failure is the favoured phrase.

Now the powerful and vocal green league has demonstrated a new intolerance of hydrocarbons by finding an avenue through which to strike at the heart of the global fossil fuel network. The new strategy seeks to eradicate the value of all energy companies' assets held in recoverable reserves of oil, gas, and coal in the ground. This is the so-called stranded assets policy which comes as the response to Lord Stern's impatient 'Why are we waiting?' question.

The argument is that half or more of the fossil fuels still buried in the ground are due to become stranded assets that can never be used and burnt. Governments across the planet, according to this thesis, will make their consumption uneconomic and maybe illegal. They are therefore valueless.

To back up this contention, campaigners call on investors – notably the biggest pension funds, sovereign wealth funds and institutional holdings, such as the Church of England or the Wellcome Trust – to demonstrate moral integrity and divest their holdings in energy company stocks.

The cause is gaining some momentum. The governor of the Bank of England, Mark Carney, has had this to say:

> 'The exposure of UK investors, including insurance companies, to these shifts is potentially huge... The challenges currently posed by climate change pale into insignificance compared with what might come. Once climate change becomes a defining issue for financial stability, it may already be too late.'[2]

Whether the governor was right to move out of high finance and monetary policy into these controversial waters is debatable,

2. Mark Carney speaking at a Lloyd's of London dinner, 28 September 2015.

but his comments highlight the flaws in the whole approach. The damage in the form of denial of investment funds to fossil reserve-owning entities could only ever affect a handful of quoted Western energy companies. Major stock-listed companies own a tiny percentage of the world's total reserves. Much the greatest proportion of the fossil fuel reserves is held by state-supported enterprises and is outside Western market reach. The attack on Western quoted companies would only constitute gesture politics, bringing with it, if the governor is right, a good deal of collateral financial and economic damage.

6. Cooperation

More importantly, it would be against the wrong target. By whom is the technology going to be driven forward – the technology that cuts the emissions from fossil fuel burning and increases efficiency in every aspect of electricity generation, transmission, and use? The answer can only be by the major energy companies of the world, the very oil, gas, mining, and power companies which the campaigners vilify. The low-carbon and greener path lies not through the destruction of energy companies and their investing power but through their innovation projects and their new investment programmes, through the vigour with which they develop and commercialize new low-carbon technologies. Cleaner methods of winning, processing, and consuming fossil fuels offer the biggest practical – and least harmful – opportunity for reducing carbon emissions.

Trying to defy the facts, closing down fossil fuels, and confining new research and cost-reducing technologies to the green sector may be noble and idealistic. But it drives straight into both political and environmental brick walls. It is not the way to solve the trilemma or produce a better, cleaner planet. It sets the two 'empires' against each other when the only hope is that they work together.

Energy Empires in Collision argues for the cooperative route – that energy policy should aim to unite the two worlds, fossil fuels and green ambitions, and put capitalism and green energy on the same side. The strategic aim should be to harness the immense power of market forces to bring about a balanced energy future and one which serves all three of the so far unattained trilemma goals – cheap power for the poor, reliable power supplies, and steady progress towards lower-carbon emissions and climate stabilization.

The green empire may claim to have the moral high ground – to be 'saving the planet' and meeting an existential threat to Gaia, as well as giving a glow of virtue to its enthusiastic adherents and leaders. But unless the green cause is aligned with fundamental economic forces, it will fail to make anything other than minor inroads into the hydrocarbon-dominated world of the past and present.

War between the two camps is fruitless and wasteful. Neither side can win completely. Both sides should work together – and must – if humanity is to prosper, if both climate violence and other scourges are to be curbed, and our global future, safeguarded.

The Paris Agreement of 2015 was described as 'a message to the marketplace'. But markets in the end react to prices and costs. For the Paris message to work on market determinants it will entail the biggest unravelling and reconstruction of the entire industrial world ever contemplated by humankind. This will be necessary even to keep within the limit of a 2.7% average increase in global temperatures (above 1990 levels) by 2050, let alone the 2% deemed by scientists as the catastrophe level, let alone the 1.6% dreamed about in the Paris Agreement of 2015.

The End Picture

There is, and can be, no end picture in a process of this kind. Technology is moving ahead so fast, and imposing such a powerful impact on human behaviour, social systems, and geopolitics that there can be no destination, no closure point, and no neat completion of the picture with global warming solved and plentiful cheap energy assured.

Analysts like to take refuge in scenarios as they peer ahead into the uncertain world of energy, with all its associated factors of politics, violent upheaval, so-called 'high-impact' events, such as terrorist outrages or natural disasters, and technological innovations. Any assumption can be invalidated overnight by unpredicted events; any projection can be bent in new directions.

Nevertheless, these things must not be allowed to blind us to some of the biggest new developments and realities in the dynamics of world energy and in the trend and range of technical possibilities.

Chief amongst these realities is that *no* forms of energy, whether from fossil fuels or newer sources, are in short supply. Huge new volumes of oil and gas are still to be developed, with technology pushing ahead all the time to transform the cost figures and open up hitherto closed areas to full economic and commercial production.

And at the same time, energy-efficient methods and innovations are racing ahead to dampen the growth of demand for energy. So we have at least one certainty – higher supply growth and lower demand growth, worldwide and continuous, which can only mean prolonged weaker oil and gas prices with all the knock-on effects described in previous parts of this chapter.

This is the new factor in the ever-unfolding saga of energy. Its impact is already being felt in dozens of areas far removed from the conventional energy scene. In Western capitals it has shifted attention on the boiling and seething Middle East away from oil and towards the new evil extruded out of schismatic Islam. In major consumer countries, it is sending a welcome boost to economic recovery from the deep wounds of 2008. In oil-producing nations it is producing new budget challenges which can rapidly turn into political challenges, some violent. In government circles it is presenting new conundrums on how to make green progress without hurt and injustice and without inevitable popular backlash.

The stories of the last two decades – that power is shifting back to Asia and the emerging world, that Europe is in decline, that the Middle East is 'finished' as an energy source, that the US has lost its way and its unipolar leadership role – themselves barely fixed in the minds of many policymakers, now all need to be revisited yet again, almost before they have been digested as the wheel of technology spins ever faster and the US takes over (again) as the world's largest energy power. Some aspects will remain valid; some are melting away.

Peering Ahead

Long-term forecasting energy developments is a fool's game. But let's try a modest joining of the jigsaw pieces for just a decade ahead.

For the next ten years, 50% of the world's oil and gas will continue to come from an arc of countries stretching from

Siberian Russia to the Persian Gulf. Every major event concerning this geographic area will send tremors through world oil markets and cause increasing price volatility.

OPEC, which has dominated the oil world for almost half a century, is no longer in a position to calm this volatility or deliver the huge assured revenue streams of the past.

Governments of producer countries in these areas will consequently be more challenged by rebel and terrorist forces, both religious and tribal, under various labels, mostly Islamic. Vladimir Putin's Russia will be checked in its ambitions and aims, especially in Ukraine. Algeria, Mexico, and Venezuela will all have to repel renewed destabilizing pressures from discontented groups within and enemies without.

Iran, the deadly foe of the Saudis, the bastion of the Shiites, will face mixed fortunes as its oil revenues are severely curbed by lower crude oil prices despite the lifting of sanctions.

The technology which has unleashed 'unconventional' oil and gas resources on a colossal scale is moving the centre of gravity of new exploration and production away from the Middle East. America's fracking boom has altered the whole pattern of world markets, although new drilling will slow as long as prices stay down in, or even below, the $45-60 range. If they go lower still, this will slow the trend away from the Middle East and give the really low-cost producers, such as Saudi Arabia, a period of remission, but the overall trend will not be halted.

Rising efficiency in energy use, and reduced intensity of energy in production all round the world, will slow down the world's growing thirst for energy and power. In the West, demand will be flat or shrinking; in the East, it will slow down somewhat. The simple and inevitable 'cause and consequence', with technology increasing supply of all kinds and reducing demand in every country, is that prices will fall in order to adjust to new market conditions.

Coal will remain by far the cheapest and therefore the most attractive fuel for the developing world. This is in flat contradiction of the aims and goals of numerous policymakers in many countries, in Europe in particular, and is the opposite of what many governments are telling the public.

Renewable energy, notably wind and solar power, will succeed or fail depending on their cost. Cheaper oil and gas will make the support needs of current renewables much greater, but will perhaps challenge their instigators to get costs down and apply new technology much faster.

This is where the best hope for a stabler energy future lies – in dramatic cost reductions in renewable energy. Some of that is happening. Because solar panels are so cheap, there is a real chance that solar power could compete directly with oil and gas and coal, and do so without subsidy.

And this is where public funds should be concentrated – to backing research and searches for the best possible new ideas in energy production from all sources, and in energy use. This is the friendliest way to assist the world's poor towards faster and more sustainable development.

Fracking enthusiasm will slow marginally in the US and round the world as the weak oil price persists. Only the really low-cost oil and gas sources will prosper, to the benefit of the best and most profitable shale producers in the USA, and so will the cheapest 'out-of-the-sand' oil from the biggest and most stable Middle Eastern producers (Saudi Arabia, Kuwait, and the UAE), although even in these countries oil production costs are rising. But where oil can still be produced for a few dollars or even a few cents, its markets will grow.

Consumers in the West and in China will receive a boost over the next few years from cheaper oil. So will Japan, although in different ways. Not only will Japan's current enormous bill for oil and gas imports fall, but if and when it goes back to nuclear power, its energy costs will fall because its nuclear stations are already built and the electricity which flows from them, once constructed and paid for, is cheap and reliable.

Poorer states, struggling to bring power to their peoples, will benefit, if allowed to, from both lower fossil fuel prices AND powerful technology heralding far greater efficiency in the generation, transmission, and use of electricity. In these societies in particular, an alliance between the fossil and green worlds will bring benefits.

Energy-consuming regions like Europe will enjoy the benefits of cheaper fuel – unless their political masters are foolish enough to try piling even more charges and taxes on oil and gas as its price falls, in an attempt to rescue current high-cost green investments. Misguided experts can already be heard or noted in the newspaper comment and letter columns advocating just this disastrous course. The only place where this kind of policy could be justified is for petrol and diesel sales, where some of the 40% drop in forecourt prices in 2015 might reasonably be diverted back to the tax authorities to help pay off heavily over-indebted national finances.

Nuclear power investment will languish despite its blessing by environmentally concerned and green authorities – unless and until dramatic cost reductions can be engineered via new technologies. Much the greatest hope is for smaller nuclear plant modules that can be factory-constructed.

These developments would redraw the map of world power and influence once again. Europe could lift itself out of its current stagnation. China could move ahead faster again despite current financial wobbles. Japan could resume its economic powerhouse status. Iraq and Iran could fall back and grow weaker (and maybe still more unstable) as their oil revenues shrink. Even as Iran starts to export more, it could find itself earning less. The very low-cost oil producers of the Middle East, the Emirates, and the Saudis – written off by some until recently – could gain a second wind as all the facilities and projects built on assumptions of an unending $80-100 dollar oil price have to be scaled down or abandoned. The US could have mixed fortunes, but with increased consumer spending generally trumping the loss of momentum with their shale revolution.

The Unending Revolution

No society, no state, no country, and no great energy company will be insulated from these changes. No side will win outright in the transforming world energy struggle. All will be forced to cooperate and to adapt. All are connected through a web of interdependent events and processes. The energy revolution is unending and continues. Those who appreciate what is actually happening, those who read the price and market signals correctly, and those who are agile enough to respond, will come through best.

The outcomes of modern technological advance are gloriously unpredictable. There is at least a chance that the familiar command and control system of power generation and distribution throughout the advanced industrial world will give way to something far more dispersed – and that this will be the pattern in the newer industrializing and developing countries as well. The dispersing and dissolving impact of the information revolution and of total world connectivity will be no less on energy industries and the whole energy landscape than it is proving to be on other industries.

Individual industrial plants may increasingly produce their own power needs. Homes may almost all become self-supplying, transforming themselves into small power plants that can store electricity and generate even higher voltage current. Transport, already vastly more efficient than it was a decade ago, will require insignificant hydrocarbon fuel. Communities may be able to generate sufficient electricity to power all civic amenities. Hospitals, schools, and all public institutions may have access to locally generated supplies. Power lines and pylons may go the same way as telephone wires and cables – into the dustbin of industrial relics.

But there can be no certainty that this is the way technology will take us. Indeed, present energy and climate policies with their inordinate and totally unrealistic price tags, as has been shown, may well frustrate progress. We just do not know what shape the energy future will take, and trying to know, trying to plan, trying to fix in detail the next energy era, is precisely what will frustrate the best possibilities.

Maybe in the end, world energy could become a dull and unsensational subject. John Maynard Keynes wrote that he hoped to see the time when the whole subject of economics took a back seat and economists were reduced to the humble status of dentists.

Unfair perhaps on dentists, but so it might one day be for the subject of energy and the society of energy actors and experts. Perhaps eventually the whole issue of providing the world with plentiful, cheap, and accessible power while effectively curbing carbon emissions, could go the same way. It could all become uninteresting, hardly worth discussing. This book has tried to explain how and why we are likely to remain so very far, and for a long while yet, from that happy, if dull, situation.

INDEX

A

Abe, Shinzo 224, 254-255
Abu Dhabi 65, 220, 251
Akita Prefecture 88
Al-Assad, President Bashar 58, 135
Al-Mazroui, Suhail 220
Al-Naimi, Ali 64
Al-Sabah, Sheikh Khalifa 23
Al-Saud, King Khalid bin Abdulaziz 55
Al-Saud, Prince Salman bin Abdulaziz 55, 56
Algeria 43, 192, 208, 276, 288, 300
Alpha Natural 86
Ambani, Mukesh 29
America
  See United States
Anderson, Keith 160
Aplin, Andrew 174
Apollo Programme 242, 291
Arab Spring 60
Asian Industrial Development Bank 88
Atlantic Ocean 280
Atlantic Trade 118, 144
Australia 39, 61, 63, 70, 87, 88, 100, 117, 152, 218, 228, 241, 279

B

Bakken 217
Baltic Sea 139
Benn, Tony 24, 150
Bevan, Aneurin 18
Bhattacharya, Sutirtha 89
Black Sea 126, 133, 185, 198
Blair, Tony 111, 150
Bloomberg Intelligence 88
Borenstein, Seth 77
Bowland Basin 173, 191
Branson, Sir Richard 29
Brent Crude 13, 39, 209, 212
Bretton Woods 279
British Columbia 254
British Gas 23
British Government 42, 60, 149, 168, 221, 258, 260, 273
British Labour 104
British Navy 213
British Treasury 23, 162, 163, 167, 260
Brussels 23, 77, 120, 124, 127, 133, 155, 274
Butler, Nick 195, 226

C

Calder Hall 102
Cameron, David 155
Canada 40, 61, 62, 113, 216, 218, 230, 254, 249, 279

Capacity Market 168
Capacity Payments 91, 166-167, 170, 171, 186-187
Carbon Capture 16, 31, 50, 93, 94, 113-115, 165, 240-241, 267
Carbon Price 160, 177
Carlton Power 170-171
Carney, Mark 296
Carington, Peter (Lord Carrington) 150
Caspian Sea 59, 125, 134, 136, 185, 197, 198, 228
Central Asia 39, 59, 63, 136, 185, 270, 275, 276, 278
Central Electricity 23, 259
Central Europe 62, 87, 124, 133, 181, 188, 189, 194, 196
China 32, 34, 39, 44, 51, 59, 61, 62, 63, 70, 75, 76, 77, 84, 86, 87-88, 89, 94-96, 97, 98, 114, 117, 122, 136, 186, 201, 202, 204, 223, 224, 232, 234, 237, 248, 250-252, 256-258, 264-283, 290, 302-303
China Development Bank 277
Churchill, Winston 126
Climate Change 42, 45, 46, 62, 63, 69, 75, 77, 78, 80, 81, 92, 95, 96, 99, 107, 150, 153, 155-156, 177, 178, 182, 193, 202, 203, 204, 234, 265, 266, 281, 290, 291, 296
Common Agricultural Policy 195
Commonwealth Office 156
Continental Shelf 190, 192, 231
Crystol Energy 21
Czech Republic 87, 116, 184, 249, 253

D

Diaoyu (or Senkaku) Islands 267, 281
Davey, Ed 150, 153
Deben, Lord 193
Deutsche Bank 227
Dudley, Bob 31

E

Eagle Ford 231
East Asia 51, 251, 278
Eastern Europe 142, 189, 277, 278
Eastern Mediterranean 61, 197, 218, 228
Electricity Market Reform 166, 178-179, 186
Elliott, Matthew 120
Emissions Trading 70, 91, 101, 118, 120, 153, 160, 181
Energiewende 99, 123, 138, 197, 253
Energy Council 236
European Council 143
European Parliament 48, 194
European Union 29, 35, 60, 69, 77, 91, 103, 117, 118, 119, 120, 121, 127, 128, 132, 141, 142, 143, 153, 155, 179, 181, 183, 186, 188, 192, 195, 196, 197, 198, 199, 276

F

Fort Murray 219, 230
Fourth Plenum 97, 224, 270
France 86, 102, 125, 128, 132, 165, 184, 187, 189, 191, 213, 247, 257, 259
Fraser, Hugh 24
Framatome 250, 258
Fukui Prefecture 255
Fukushima 43, 50, 85, 117, 123, 139, 245-247, 252, 253, 254-255, 287

G

Gabriel, Sigmar 89, 100
Gates, Bill, 29, 243
Gatliff, Robert 174
Germany 34, 48, 50, 85-86, 89-90, 99, 100, 102, 103, 116, 117, 123, 125, 131, 137, 138-141, 166, 183, 187, 196, 197, 204, 247, 252, 253, 272, 285, 291, 294
Gilmer, Allen 230
Global Warming 28, 42, 44, 47, 74, 80, 81, 86, 92, 181, 182, 200, 237, 284, 298
Gore, Al 155, 201
Goyal, Piyush 89
Grantham Research 92
Great Britain
  see United Kingdom
Green Capitalism 295-296
Guofeng, Hua 264, 265
Gulf States 23, 47, 222, 276

H

Heath, Edward 150, 192
Helm, Dieter 16-17, 19, 123, 286, 287, 289, 290
Hinkley Point 51, 168, 197, 250, 256-258, 261, 277
Holyrood Government 177
Hong Kong 267-268
Howell, David 14, 150, 168, 201
Huhne, Chris 150, 153, 156
Hussein, Saddam 208, 217
Hyde Park 264

I

India 34, 44, 63, 70, 75, 86, 87, 88, 89, 94, 95, 96, 112, 117, 118, 120, 122, 136, 204, 223, 249, 251, 252, 279, 290
Indian Ocean 280
Indonesia 86, 88, 251
Industrial Revolution 71, 129, 144, 146, 174, 176
International Energy Agency 56, 68, 71, 113
Ingham, Bernard 23
Islamic State (Daesh) 43, 58, 60, 218, 229
Israel 40, 60, 136, 197, 218, 228

J

Jacques, Martin 272, 280

Japan 41, 43, 50, 51, 63, 77, 86, 88, 89, 98-99, 117, 139, 173, 217, 224, 245-247, 252, 253, 254-255, 256, 267, 279, 302, 303
Jinping, Xi 257, 270

K

Kaczynski , President Jarosław 144
Kansai Electric 88
Kensington Palace 264
Keqiang, Li 277, 278
Keynes, John Maynard 304
Klein, Naomi 44, 269
Kyoto Protocol 101, 202

L

Labour Government 20, 24, 105, 150, 154, 156
Latin America 39, 62, 72, 79, 223, 236, 276
Law and Justice Party 144
Lawson, Nigel 150
Liberal Democrats 155, 178
Libya 40, 43, 60, 208, 226, 288
Liquefied Gas 99, 110, 125, 131, 132, 134, 135, 136, 143, 158, 173, 188-189, 192, 196, 218, 224, 270, 274, 275, 276, 280
Lisbon Treaty 120, 121
Lovelock, James 50, 83, 247, 252

M

Macleod, Iain 182
Merkel, Angela 85, 89, 139, 140, 252
Marubeni Corporation 88
Mecca 55
Middle East 28, 34, 43, 47, 51, 53, 57, 60, 62, 63, 64, 67, 68, 70, 71, 74, 84, 98, 120, 184, 192, 199, 208, 212-215, 217, 222, 224, 265, 267, 281, 283, 299, 300, 301, 302, 303
Miliband, Ed 150, 152
Mitsubishi Electric 243
Mitsubishi Heavy Industries 256
Modi, Narendra 63, 96
Moore, Henry 137

N

Nabucco 134, 185, 186, 198
Nakhle, Dr. Carole 14, 21, 201, 233
National Coal Board 23
National Grid 146, 147, 149, 157-159, 171-172, 177
Neslen, Arthur 77
New Silk Road 278
New York 29, 59
Nigeria 40, 41, 42, 43, 229, 276
North Africa 84, 192, 197, 213
North Sea 22, 40, 62, 110, 129, 139, 146, 150, 166, 173, 177, 192, 197, 217, 219, 231-232, 264
Norway 23, 125, 165, 187, 192, 197, 228, 240, 272, 273
Nuclear Power 32, 34, 41, 43, 48, 49, 50-51, 63, 71, 73, 83-84,

84-85, 88, 89, 90, 97, 99, 102, 110, 111, 114, 115, 117, 118, 123, 125, 128, 139, 140, 147-148, 158, 168, 171, 172, 178, 179, 181, 183-184, 187, 195, 197, 201, 205-206, 210, 215, 224, 229, 234, 235, 236, 237, 243, 244, 245-263, 266, 271, 273, 274, 287, 302, 303
Nurse, Sir Paul 80

O

Obama, Barack 86, 95, 97, 271
Oettinger, Günther 77
Orban, Viktor 277
Organization of Petroleum Exporting Countries (OPEC) 35, 40, 41, 47, 59, 61, 63, 64, 67, 68, 71, 184, 212-214, 220-223, 226, 227-228, 288, 300

P

Papua New Guinea 88
Paris Agreement 17, 29, 47, 285, 298
Parkinson, Cecil 150
Paterson, Owen 104
Peabody Energy 86
Peiser, Benny 181, 182
Persian Gulf 300
Poland 49, 103, 122, 131, 141-144, 145, 184, 191, 252, 263, 277
Powell, Enoch 182
Putin, Vladimir 48, 124, 131, 133, 135, 138, 185, 215, 230, 300

Q

Queen Elizabeth 102

R

Rampton, Sir Jack 23
Repsol Brasil 277
Riyadh 56
Rooke, Sir Denis 23
Royal Society 80, 105
Rudd, Amber 86
Rühl, Christof 122, 231
Ruhrgas 131
Russell, Bertrand 154
Russia 34, 40, 41, 44, 45, 48, 51, 58, 63, 70, 77, 79, 87, 95, 96, 98, 100, 111, 116, 120, 124, 125, 126-127, 128, 129, 130-137, 137-138, 140, 142, 143, 145, 152, 178, 181, 185-186, 188, 190, 192, 195, 196, 198-199, 201, 204, 215, 217, 218-219, 220-221, 226, 227, 228, 230, 232, 248-249, 251, 253, 258, 263, 266, 275-276, 290, 294, 300

S

Sachs, Jeffrey 42
Sata, Michael 276

Saudi Arabia 54-58, 64-65, 68, 98, 120, 190, 212, 214, 217, 221, 226, 227, 229, 266, 276, 301, 302
Scargill, Arthur 22, 260, 287
Schlesinger, Jim 23
Schmidt, Helmut 137
Scotland 14, 161, 177
Scottish Power 160
Šefcovic, Maroš 198
Senate Committee 181-182
Senkaku (or Diaoyu) Islands 267, 281
Shah Deniz 136
Shale 34, 39, 41, 47, 61, 62, 70, 73, 74, 77, 86, 87, 92, 117, 123, 124, 126, 132, 135, 136, 141, 142, 143, 173-177, 189-192, 206, 210, 213, 214, 216, 217, 219, 221, 227, 230-231, 232, 253, 254, 266, 267, 270, 274-275, 277, 302, 303
Shell 31, 32, 70, 211, 215, 219
Single Market 121, 127, 181, 188
Soros, George 29, 87, 90
South Africa 88, 249
South Korea 77, 88, 249
South Stream 126, 133, 185
Stern Review 39, 300
Stern, Lord 39, 45, 200, 295
Summers, Lawrence 279

T

Tajani, Antonio 48
Tempest, Paul 20
Tesla Motors 165
Texas Intermediate 39, 209
Thar Coal 88
Thatcher, Margaret 23, 24, 137, 184, 260
Tiafen, Liu 274
Tokyo Electric Power Company 246, 255
Trans-Adriatic Pipeline 198
Trans-Anatolian Pipeline 16
Tuchman, Barbara 199
Turkey 43, 60, 133, 134, 136, 185, 186, 198, 199, 218
Tusk, Donald 142, 143, 188

U

Ukraine 61, 111, 124, 126, 130-132, 133, 135, 138, 142, 155, 172, 178, 185, 199, 215, 249, 300
United Arab Emirates 120, 214, 220, 221, 227, 249, 258, 302
United Kingdom 140, 144, 145, 153
United Nations 29, 30, 63, 75, 78
United States 34, 38, 56, 74, 77, 86, 87, 90, 117, 124, 125, 134, 135, 140, 144, 164, 172, 175, 181, 182, 189, 201, 202, 213, 217, 218, 219, 221, 230, 237, 241, 253, 263, 266, 267, 268, 271, 272, 274, 275, 279, 290, 294, 299, 300, 302, 303

V

Varley, John 150

Viking Electricity 186

W

Wakeham, John 150
Wales 165, 177
Walker, Ian 20
Walker, Peter 150
Ward, Bob 92
Weald Region 174
Wellcome Trust 295
West Africa 61, 172
Western Europe 131, 132, 133, 134, 141, 185, 188, 204, 215, 278
Windsor Energy Group 20
Wolf Camp 231
World Bank 88
World Health Organization 267
World Nuclear Association 249, 250, 262

X

Xinjiang Province 279

Y

Yamani, Sheikh Ahmed Zaki 23, 54, 55, 57, 58
Yergin, Daniel 16

Z

Zhao, Wang 278
Zuckerberg, Mark 29